Fractional Calculus and Special Functions with Applications

Fractional Calculus and Special Functions with Applications

Editors

Mehmet Ali Özarslan
Arran Fernandez
Ivan Area

MDPI • Basel • Beijing • Wuhan • Barcelona • Belgrade • Manchester • Tokyo • Cluj • Tianjin

Editors
Mehmet Ali Özarslan
Department of Mathematics,
Faculty of Arts and Sciences,
Eastern Mediterranean
University
Northern Cyprus, via Mersin 10
Turkey

Arran Fernandez
Department of Mathematics,
Faculty of Arts and Sciences,
Eastern Mediterranean
University
Northern Cyprus, via Mersin 10
Turkey

Ivan Area
CITMAga, Universidade de
Vigo, Departamento de
Matemática Aplicada II, E.E.
Aeronáutica e do Espazo,
Campus As Lagoas-Ourense
Spain

Editorial Office
MDPI
St. Alban-Anlage 66
4052 Basel, Switzerland

This is a reprint of articles from the Special Issue published online in the open access journal *Fractal and Fractional* (ISSN 2504-3110) (available at: https://www.mdpi.com/journal/fractalfract/special_issues/Fract_Calculus).

For citation purposes, cite each article independently as indicated on the article page online and as indicated below:

LastName, A.A.; LastName, B.B.; LastName, C.C. Article Title. *Journal Name* **Year**, *Volume Number*, Page Range.

ISBN 978-3-0365-3617-0 (Hbk)
ISBN 978-3-0365-3618-7 (PDF)

© 2022 by the authors. Articles in this book are Open Access and distributed under the Creative Commons Attribution (CC BY) license, which allows users to download, copy and build upon published articles, as long as the author and publisher are properly credited, which ensures maximum dissemination and a wider impact of our publications.

The book as a whole is distributed by MDPI under the terms and conditions of the Creative Commons license CC BY-NC-ND.

Contents

About the Editors vii

Mehmet Ali Özarslan, Arran Fernandez, Iván Area
Editorial for Special Issue "Fractional Calculus and Special Functions with Applications"
Reprinted from: *Fractal Fract.* **2021**, *5*, 224, doi:10.3390/fractalfract5040224 1

Shorog Aljoudi, Bashir Ahmad and Ahmed Alsaedi
Existence and Uniqueness Results for a Coupled System of Caputo-Hadamard Fractional Differential Equations with Nonlocal Hadamard Type Integral Boundary Conditions
Reprinted from: *Fractal Fract.* **2020**, *4*, 13, doi:10.3390/fractalfract4020013 5

Ahmed Salem and Balqees Alghamdi
Multi-Strip and Multi-Point Boundary Conditions for Fractional Langevin Equation
Reprinted from: *Fractal Fract.* **2020**, *4*, 18, doi:10.3390/fractalfract4020018 21

Houssine Zine and Delfim F. M. Torres
A Stochastic Fractional Calculus with Applications to Variational Principles
Reprinted from: *Fractal Fract.* **2020**, *4*, 38, doi:10.3390/fractalfract4030038 35

Arran Fernandez and Iftikhar Husain
Modified Mittag-Leffler Functions with Applications in Complex Formulae for Fractional Calculus
Reprinted from: *Fractal Fract.* **2020**, *4*, 45, doi:10.3390/fractalfract4030045 47

Övgü Gürel Yılmaz, Rabia Aktaş and Fatma Taşdelen
On Some Formulas for the k-Analogue of Appell Functions and Generating Relations via k-Fractional Derivative
Reprinted from: *Fractal Fract.* **2020**, *4*, 48, doi:10.3390/fractalfract4040048 63

Bahar Acay and Mustafa Inc
Electrical Circuits RC, LC, and RLC under Generalized Type Non-Local Singular Fractional Operator
Reprinted from: *Fractal Fract.* **2021**, *5*, 9, doi:10.3390/fractalfract5010009 83

Sümeyra Uçar, Esmehan Uçar, Fırat Evirgen and Necati Özdemir
A Fractional SAIDR Model in the Frame of Atangana–Baleanu Derivative
Reprinted from: *Fractal Fract.* **2021**, *5*, 32, doi:10.3390/fractalfract5020032 101

Mehmet Ali Özarslan and Arran Fernandez
On a Five-Parameter Mittag-Leffler Function and the Corresponding Bivariate Fractional Operators
Reprinted from: *Fractal Fract.* **2021**, *5*, 45, doi:10.3390/fractalfract5020045 119

Esmail Bargamadi, Kazem Nouri, Leila Torkzadeh, Amin Jajarmi
Solving a System of Fractional-Order Volterra-Fredholm Integro-Differential Equations with Weakly Singular Kernels via the Second Chebyshev Wavelets Method
Reprinted from: *Fractal Fract.* **2021**, *5*, 70, doi:10.3390/fractalfract5030070 141

About the Editors

Mehmet Ali Özarslan is a Professor at the Department of Mathematics, Eastern Mediterranean University, Northern Cyprus. He has published over 100 articles in reputable journals. These publications are especially focused on special functions, approximation theory, and fractional calculus. His current research is specifically related to bivariate fractional calculus operators constructed using general special functions including Mittag-Leffler functions which have semigroup properties and general Appell polynomials.

Arran Fernandez is an Assistant Professor at the Department of Mathematics, Eastern Mediterranean University, Northern Cyprus. He studied at the University of Cambridge, where he was the youngest ever Senior Wrangler. He has published over 50 research papers in reputable international journals. His research interests lie in fractional calculus, fractional differential equations, asymptotic analysis, and special functions including zeta functions and Mittag-Leffler functions.

Ivan Area is a Professor at the Departamento de Matemática Aplicada II, E.E. Aeronáutica e do Espazo Universidade de Vigo, Spain. He has published over 100 articles related to orthogonal polynomials and special functions in leading international journals. His recent research has also focused on fractional analysis and bioinformatics.

Editorial

Editorial for Special Issue "Fractional Calculus and Special Functions with Applications"

Mehmet Ali Özarslan [1,*], Arran Fernandez [1] and Iván Area [2]

[1] Department of Mathematics, Faculty of Arts and Sciences, Eastern Mediterranean University, 99628 Famagusta, Northern Cyprus, Via Mersin-10, Turkey; arran.fernandez@emu.edu.tr
[2] CITMAga, Universidade de Vigo, Departamento de Matemática Aplicada II, E.E. Aeronáutica e do Espazo, Campus As Lagoas-Ourense, 32004 Ourense, Spain; area@uvigo.gal
* Correspondence: mehmetali.ozarslan@emu.edu.tr

Citation: Özarslan, M.A.; Fernandez, A.; Area, I. Editorial for Special Issue "Fractional Calculus and Special Functions with Applications". *Fractal Fract.* **2021**, *5*, 224. https://doi.org/10.3390/fractalfract5040224

Received: 2 November 2021
Accepted: 10 November 2021
Published: 17 November 2021

Publisher's Note: MDPI stays neutral with regard to jurisdictional claims in published maps and institutional affiliations.

Copyright: © 2021 by the authors. Licensee MDPI, Basel, Switzerland. This article is an open access article distributed under the terms and conditions of the Creative Commons Attribution (CC BY) license (https://creativecommons.org/licenses/by/4.0/).

The study of fractional integrals and fractional derivatives has a long history, and they have many real-world applications due to their properties of interpolation between operators of integer order. This field has covered classical fractional operators such as Riemann–Liouville, Weyl, Caputo, Grünwald–Letnikov, etc. Also, especially in the last two decades, many new operators have appeared, often defined using integrals with special functions in the kernel, such as Atangana–Baleanu, Prabhakar, Marichev–Saigo–Maeda, and tempered, as well as their extended or multivariable forms. These have been intensively studied because they can also be useful in modelling and analysing real-world processes because of their different properties and behaviours, which are comparable to those of the classical operators.

Special functions, such as the Mittag-Leffler functions, hypergeometric functions, Fox's H-functions, Wright functions, Bessel and hyper-Bessel functions, etc., also have some more classical and fundamental connections with fractional calculus. Some of them, such as the Mittag-Leffler function and its generalisations, appear naturally as solutions of fractional differential equations or fractional difference equations. Furthermore, many interesting relationships between different special functions may be discovered using the operators of fractional calculus. Certain special functions have also been applied to analyse the qualitative properties of fractional differential equations, such as the concept of Mittag-Leffler stability.

In early 2020, we opened a Special Issue in the journal *Fractal and Fractional*, with the aim of exploring and celebrating the diverse connections between fractional calculus and special functions, as well as their associated applications. The deadline was initially set as 31 December 2020, and was later extended to 31 March 2021 after the havoc caused by the COVID-19 pandemic. We received a total of 15 submissions for this Special Issue and, after a thorough peer-review process, nine of these were ultimately published, including several from experts in the field whom we had personally invited to contribute.

The published papers in our Special Issue are briefly summarised as follows.

In [1], Aljoudi et al. considered a nonlinear coupled system of Caputo–Hadamard fractional ordinary differential equations on a finite interval, with Hadamard integral boundary conditions and incommensurate fractional orders between 1 and 2. Under certain boundedness and Lipschitz-type assumptions, they proved the existence of solutions for this system, and the uniqueness of solutions under some extra boundedness assumptions.

In [2], Salem and Alghamdi considered a nonlinear sequential-type Caputo fractional ordinary differential equation on a finite interval, with nonlocal multi-point boundary conditions and an overall fractional order between 1 and 3. They proved the existence and/or uniqueness of solutions for this Langevin-type equation in three main results, under an array of possible conditions.

In [3], Zine and Torres introduced a new type of stochastic fractional operator, a way of applying fractional integrals and derivatives to stochastic processes. They proved many fundamental properties of these operators, including boundedness, semigroup and inversion properties, and an integration by parts rule, before posing stochastic fractional Euler–Lagrange equations to investigate the variational principles of the new operators.

In [4], Fernandez and Husain defined and investigated modified versions of the classical Mittag-Leffler functions of one, two, and three parameters. They found appropriate convergence conditions for the new series in each case, established complex integral representations of the new functions, and then used them to extend the definition of Atangana–Baleanu and Prabhakar fractional calculus, providing analytic continuations of the original definitions to wider domains for the parameter α.

In [5], Yılmaz et al. studied k-generalised Appell functions, based on the existing theory of k-fractional calculus and k-variants of special functions such as the gamma, beta, and hypergeometric functions (the k-variants being identical to the original versions up to some substitutions and constant multiples). They proved various functional equation relations and generating relations for the k-generalised Appell functions using k-fractional derivatives. Please kindly note that a corrigendum to this paper was also published [5].

In [6], Acay and İnç studied several variants of a differential equation used to model RC, LC, and RLC electric circuits under Kirchhoff's law. The function representing the source voltage was taken to be either constant, exponential, or a power function, and the differential operator was taken to be a so-called non-local fractional M-derivative, which is a constant times the usual Caputo derivative taken with respect to a power function. A comparative analysis was performed to compare the results achieved by using the M-derivative and by using the usual Caputo derivative with respect to t.

In [7], Uçar et al. considered a system of first-order ordinary differential equations, which is used to model the effect of computer worms, and replaced the first-order derivatives with fractional derivatives of Atangana–Baleanu type to obtain a different system, which they studied using fixed-point and Laplace transform techniques to prove existence, uniqueness, and stability properties.

In [8], Özarslan and Fernandez introduced a new five-parameter Mittag-Leffler function, defined by a single series but used to construct bivariate (double integral) fractional operators. They proved fundamental properties such as boundedness, Laplace transforms, semigroup and inversion properties, and series formulae. In the non-singular case, they derived a special case which is a mixed bivariate version of the Atangana–Baleanu operators.

In [9], Bargamadi et al. considered a coupled system of integro-differential equations involving Caputo fractional derivatives, with simple initial conditions and incommensurate fractional orders between 0 and 1. They used the Chebyshev wavelets method to estimate the Caputo derivatives and find approximate numerical solutions to the system, and performed error analysis on their approximations both analytically and numerically.

As the handling Guest Editors, we would like to express our gratitude to all authors for their contributions, as well as to all the peer reviewers who helped to improve the quality of the submissions. We would also like to thank Ms. Jingjing Yang from the journal office for her prompt assistance throughout the process of managing this Special Issue.

Funding: This research received no external funding.

Conflicts of Interest: The authors declare no conflict of interest.

References

1. Aljoudi, S.; Ahmad, B.; Alsaedi, A. Existence and Uniqueness Results for a Coupled System of Caputo—Hadamard Fractional Differential Equations with Nonlocal Hadamard Type Boundary Conditions. *Fractal Fract.* **2020**, *4*, 13. [CrossRef]
2. Salem, A.; Alghamdi, B. Multi-Strip and Multi-Point Boundary Conditions for Fractional Langevin Equation. *Fractal Fract.* **2020**, *4*, 18. [CrossRef]
3. Zine, H.; Torres, D.F.M. A Stochastic Fractional Calculus with Applications to Variational Principles. *Fractal Fract.* **2020**, *4*, 38. [CrossRef]
4. Fernandez, A.; Husain, I. Modified Mittag-Leffler Functions with Applications in Complex Formulae for Fractional Calculus. *Fractal Fract.* **2020**, *4*, 45. [CrossRef]

5. Gürel Yılmaz, Ö.; Aktaş, R.; Taşdelen, F. On Some Formulas for the k-Analogue of Appell Functions and Generating Relations via k-Fractional Derivative. *Fractal Fract.* **2020**, *4*, 48; Erratum in **2020**, *4*, 60. [CrossRef]
6. Acay, B.; Inç, M. Electrical Circuits RC, LC, and RLC under Generalized Type Non-Local Singular Fractional Operator. *Fractal Fract.* **2021**, *5*, 9. [CrossRef]
7. Uçar, E.; Uçar, S.; Evirgen, F.; Özdemir, N. A Fractional SAIDR Model in the Frame of Atangana–Baleanu Derivative. *Fractal Fract.* **2021**, *5*, 32. [CrossRef]
8. Özarslan, M.A.; Fernandez, A. On a Five-Parameter Mittag-Leffler Function and the Corresponding Bivariate Fractional Operators. *Fractal Fract.* **2021**, *5*, 45. [CrossRef]
9. Bargamadi, E.; Torkzadeh, L.; Nouri, K.; Jajarmi, A. Solving a System of Fractional-Order Volterra–Fredholm Integro-Differential Equations with Weakly Singular Kernels via the Second Chebyshev Wavelets Method. *Fractal Fract.* **2021**, *5*, 70. [CrossRef]

Article

Existence and Uniqueness Results for a Coupled System of Caputo-Hadamard Fractional Differential Equations with Nonlocal Hadamard Type Integral Boundary Conditions

Shorog Aljoudi [1], Bashir Ahmad [2,*] and Ahmed Alsaedi [2]

[1] Department of Mathematics and Statistics, Taif University, P.O. Box 888, Taif 21974, Saudi Arabia; sh-aljoudi@hotmail.com
[2] Nonlinear Analysis and Applied Mathematics (NAAM)-Research Group, Department of Mathematics, Faculty of Science, King Abdulaziz University, P.O. Box 80203, Jeddah 21589, Saudi Arabia; aalsaedi@kau.edu.sa
* Correspondence: bahmad@kau.edu.sa

Received: 3 March 2020; Accepted: 2 April 2020; Published: 12 April 2020

Abstract: In this paper, we study a coupled system of Caputo-Hadamard type sequential fractional differential equations supplemented with nonlocal boundary conditions involving Hadamard fractional integrals. The sufficient criteria ensuring the existence and uniqueness of solutions for the given problem are obtained. We make use of the Leray-Schauder alternative and contraction mapping principle to derive the desired results. Illustrative examples for the main results are also presented.

Keywords: Caputo-Hadamard fractional derivative; coupled system; Hadamard fractional integral; boundary conditions; existence

MSC: 34A08, 34B10, 34B15

1. Introduction

Fractional calculus has emerged as an important area of investigation in view of its extensive applications in mathematical modeling of many complex and nonlocal nonlinear systems. An important characteristic of fractional-order operators is their nonlocal nature that accounts for the hereditary properties of the underlying phenomena. The interactions among macromolecules in the damping phenomenon give rise to a macroscopic stress-strain relation in terms of fractional differential operators. For the fractional law dealing with the viscoelastic materials, see [1] and the references cited therein. In [2], transport processes influenced by the past and present histories are described by the Caputo power law. For the details on dynamic memory involved in the economic processes, see [3,4]. In 1892, Hadamard [5] suggested a concept of fractional integro-differentiation in terms of the fractional power of the type $(t\frac{d}{dt})^q$ in contrast to its Riemann-Liouville counterpart of the form $(\frac{d}{dt})^q$. The Hadamard fractional derivative contains a logarithmic function of an arbitrary exponent in the kernel of the integral appearing in its definition. For the details of Hadamard fractional calculus, we refer the reader to the works [6–9]. Fractional differential equations involving Hadamard derivative attracted significant attention in recent years, for instance, see [10–20] and the references cited therein.

More recently, Jarad et al. [21] introduced Caputo modification of Hadamard fractional derivative which is more suitable for physically interpretable initial conditions as in case of Caputo fractional differential equations. One can find some recent results on Caputo-Hadamard type fractional differential equations in [22–28] and the references cited therein.

In this paper, we introduce a new class of boundary value problems consisting of Caputo-Hadamard type fractional differential equations and Hadamard type fractional integral boundary conditions. In precise terms, we investigate the following boundary value problem:

$$\begin{cases} (^C D^\alpha + \lambda {}^C D^{\alpha-1}) u(t) = f(t, u(t), v(t), {}^C D^{\bar{\xi}} v(t)), \ 1 < \alpha \le 2, \ 0 < \bar{\xi} < 1, \ \lambda > 0, \\ (^C D^\beta + \lambda {}^C D^{\beta-1}) v(t) = g(t, u(t), {}^C D^{\bar{\xi}} u(t), v(t)), \ 1 < \beta \le 2, \ 0 < \bar{\xi} < 1, \\ u(1) = 0, \ a_1 \mathcal{I}^{\gamma_1} v(\eta_1) + b_1 u(T) = K_1, \ \gamma_1 > 0, \ 1 < \eta_1 < T, \\ v(1) = 0, \ a_2 \mathcal{I}^{\gamma_2} u(\eta_2) + b_2 v(T) = K_2, \ \gamma_2 > 0, \ 1 < \eta_2 < T, \end{cases} \quad (1)$$

where $^C D^{(\cdot)}$ and $\mathcal{I}^{(\cdot)}$ respectively denote the Caputo-Hadamard fractional derivative and Hadamard fractional integral (to be defined later), $f, g : [1, T] \times \mathbb{R}^3 \to \mathbb{R}$ are given appropriate functions and $a_i, b_i, K_i, (i = 1, 2)$ are real constants.

The rest of the paper is organized as follows. In Section 2, we recall the background material related to the topic under investigation and prove an auxiliary lemma which plays a key role in deriving the desired results. Section 3 contains the main results.

2. Preliminaries

In this section, we recall some preliminary concepts of Hadamard and Caputo-Hadamard fractional calculus related to our work. We also prove an auxiliary lemma, which plays a key role in converting the given problem into a fixed point problem.

Definition 1 ([6,7]). *The Hadamard fractional integral of order $q \in \mathbb{C}$, $\mathcal{R}(q) > 0$, for a function $g \in L^p[a, b]$, $0 \le a \le t \le b \le \infty$, is defined as*

$$I_{a+}^q g(t) = \frac{1}{\Gamma(q)} \int_a^t \left(\log \frac{t}{s} \right)^{q-1} \frac{g(s)}{s} ds,$$

$$I_{b-}^q g(t) = \frac{1}{\Gamma(q)} \int_t^b \left(\log \frac{s}{t} \right)^{q-1} \frac{g(s)}{s} ds.$$

Definition 2 ([6,7]). *Let $[a, b] \subset \mathbb{R}$, $\delta = t \frac{d}{dt}$ and $AC_\delta^n[a,b] = \{g : [a,b] \to \mathbb{R} : \delta^{n-1}(g(t)) \in AC[a,b]\}$. The Hadamard derivative of fractional order q for a function $g \in AC_\delta^n[a,b]$ is defined as*

$$D_{a+}^q g(t) = \delta^n (I_{a+}^{n-q})(t) = \frac{1}{\Gamma(n-q)} \left(t \frac{d}{dt} \right)^n \int_a^t \left(\log \frac{t}{s} \right)^{n-q-1} \frac{g(s)}{s} ds,$$

$$D_{b-}^q g(t) = (-\delta)^n (I_{b-}^{n-q})(t) = \frac{1}{\Gamma(n-q)} \left(-t \frac{d}{dt} \right)^n \int_t^b \left(\log \frac{s}{t} \right)^{n-q-1} \frac{g(s)}{s} ds,$$

where $n - 1 < q < n$, $n = [q] + 1$ and $[q]$ denotes the integer part of the real number q and $\log(\cdot) = \log_e(\cdot)$.

Definition 3 ([21]). *For $\mathcal{R}(q) > 0$, $n = [\mathcal{R}(q)] + 1$, and $g \in AC_\delta^n[a,b] \ 0 \le a \le t \le b \le \infty$, the Caputo-type modification of the Hadamard fractional derivative is defined by*

$$^C D_{a+}^q g(t) = D_{a+}^q \left[g(s) - \sum_{k=0}^{n-1} \frac{\delta^k g(a)}{k!} \left(\log \frac{s}{a} \right)^k \right](t),$$

$$^C D_{b-}^q g(t) = D_{b-}^q \left[g(s) - \sum_{k=0}^{n-1} \frac{(-1)^k \delta^k g(b)}{k!} \left(\log \frac{b}{s} \right)^k \right](t).$$

Theorem 1 ([21]). *Let $\mathcal{R}(q) \geq 0, n = [\mathcal{R}(q)] + 1$ and $g \in AC^n_\delta[a,b]$, $0 \leq a \leq t \leq b \leq \infty$. Then ${}^C\mathcal{D}^q_{a+}g(t)$ and ${}^C\mathcal{D}^q_{b-}g(t)$ exist everywhere on $[a,b]$ and*

(a) *if $q \notin \mathbb{N}_0$,*

$$
{}^C\mathcal{D}^q_{a+}g(t) = \frac{1}{\Gamma(n-q)} \int_a^t \left(\log\frac{t}{s}\right)^{n-q-1} \delta^n g(s) \frac{ds}{s} = (I^{n-q}_{a+}) \delta^n g(t),
$$

$$
{}^C\mathcal{D}^q_{b-}g(t) = \frac{(-1)^n}{\Gamma(n-q)} \int_t^b \left(\log\frac{s}{t}\right)^{n-q-1} \delta^n g(s) \frac{ds}{s} = (-1)^n (I^{n-q}_{b-}) \delta^n g(t);
$$

(b) *if $q = n \in \mathbb{N}_0$,*

$$
{}^C\mathcal{D}^q_{a+}g(t) = \delta^n g(t), \qquad {}^C\mathcal{D}^q_{b-}g(t) = (-1)^n \delta^n g(t).
$$

In particular,

$$
{}^C\mathcal{D}^0_{a+}g(t) = {}^C\mathcal{D}^0_{b-}g(t) = g(t).
$$

Remark 1 ([29]). *For $q \in \mathbb{C}$ such that $0 < q < 1$, the Caputo-Hadamard fractional derivative is defined as*

$$
{}^C\mathcal{D}^q_{a+}g(t) = \frac{1}{\Gamma(1-q)} \int_a^t \left(\log\frac{t}{s}\right)^{-q} g'(s)ds,
$$

$$
{}^C\mathcal{D}^q_{b-}g(t) = \frac{-1}{\Gamma(1-q)} \int_t^b \left(\log\frac{s}{t}\right)^{-q} g'(s)ds.
$$

Lemma 1 ([21]). *Let $\mathcal{R}(q) \geq 0, n = [\mathcal{R}(q)] + 1$ and $g \in C[a,b]$. If $\mathcal{R}(q) \neq 0$ or $q \in \mathbb{N}$, then*

$$
{}^C\mathcal{D}^q_{a+}(I^q_{a+}g)(t) = g(t), \quad {}^C\mathcal{D}^q_{b-}(I^q_{b-}g)(t) = g(t).
$$

Lemma 2 ([21]). *Let $g \in AC^n_\delta[a,b]$ or $C^n_\delta[a,b]$ and $q \in \mathbb{C}$, then*

$$
I^q_{a+}({}^C\mathcal{D}^q_{a+}g)(t) = g(t) - \sum_{k=0}^{n-1} \frac{\delta^k g(a)}{k!} \left(\log\frac{t}{a}\right)^k,
$$

$$
I^q_{b-}({}^C\mathcal{D}^q_{b-}g)(t) = g(t) - \sum_{k=0}^{n-1} \frac{\delta^k g(b)}{k!} \left(\log\frac{b}{t}\right)^k.
$$

Now we present an auxiliary lemma dealing with the linear variant of the problem (1).

Lemma 3. *Let $h_1, h_2 \in AC^n_\delta[1,T]$. Then the solution of the linear system of fractional differential equations:*

$$
\begin{aligned}
({}^C\mathcal{D}^\alpha + \lambda {}^C\mathcal{D}^{\alpha-1})u(t) &= h_1(t), \\
({}^C\mathcal{D}^\beta + \lambda {}^C\mathcal{D}^{\beta-1})v(t) &= h_2(t),
\end{aligned} \qquad (2)
$$

supplemented with the boundary conditions:

$$
\begin{aligned}
u(1) &= 0, \quad a_1 \mathcal{I}^{\gamma_1} v(\eta_1) + b_1 u(T) = K_1, \; \gamma_1 > 0, \; 1 < \eta_1 < T, \\
v(1) &= 0, \quad a_2 \mathcal{I}^{\gamma_2} u(\eta_2) + b_2 v(T) = K_2, \; \gamma_2 > 0, \; 1 < \eta_2 < T,
\end{aligned} \qquad (3)
$$

is given by

$$
\begin{aligned}
u(t) &= \frac{(1-t^{-\lambda})}{\lambda \Delta} \left\{ (K_2 A_2 - K_1 B_2) + T^{-\lambda} \left[b_1 B_2 \int_1^T s^{\lambda-1} \mathcal{I}^{\alpha-1} h_1(s) ds - b_2 A_2 \int_1^T s^{\lambda-1} \mathcal{I}^{\beta-1} h_2(s) ds \right] \right. \\
&\quad + \frac{a_1 B_2}{\Gamma(\gamma_1)} \int_1^{\eta_1} \left(\log\frac{\eta_1}{s}\right)^{\gamma_1-1} s^{-(\lambda+1)} \left(\int_1^s m^{\lambda-1} \mathcal{I}^{\beta-1} h_2(m) dm \right) ds
\end{aligned}
$$

$$- \frac{a_2 A_2}{\Gamma(\gamma_2)} \int_1^{\eta_2} \left(\log \frac{\eta_2}{s}\right)^{\gamma_2-1} s^{-(\lambda+1)} \left(\int_1^s m^{\lambda-1} \mathcal{I}^{\alpha-1} h_1(m) dm\right) ds \bigg\}$$

$$+ t^{-\lambda} \int_1^t s^{\lambda-1} \mathcal{I}^{\alpha-1} h_1(s) ds, \tag{4}$$

and

$$v(t) = \frac{(1-t^{-\lambda})}{\lambda \Delta} \bigg\{ (K_1 B_1 - K_2 A_1) + T^{-\lambda} \left[b_2 A_1 \int_1^T s^{\lambda-1} \mathcal{I}^{\beta-1} h_2(s) ds - b_1 B_1 \int_1^T s^{\lambda-1} \mathcal{I}^{\alpha-1} h_1(s) ds \right]$$

$$+ \frac{a_2 A_1}{\Gamma(\gamma_2)} \int_1^{\eta_2} \left(\log \frac{\eta_2}{s}\right)^{\gamma_2-1} s^{-(\lambda+1)} \left(\int_1^s m^{\lambda-1} \mathcal{I}^{\alpha-1} h_1(m) dm\right) ds$$

$$- \frac{a_1 B_1}{\Gamma(\gamma_1)} \int_1^{\eta_1} \left(\log \frac{\eta_1}{s}\right)^{\gamma_1-1} s^{-(\lambda+1)} \left(\int_1^s m^{\lambda-1} \mathcal{I}^{\beta-1} h_2(m) dm\right) ds \bigg\}$$

$$+ t^{-\lambda} \int_1^t s^{\lambda-1} \mathcal{I}^{\beta-1} h_2(s) ds. \tag{5}$$

where

$$\Delta = B_1 A_2 - A_1 B_2 \neq 0, \tag{6}$$

$$A_1 = \frac{b_1}{\lambda}(1 - T^{-\lambda}), \quad A_2 = \frac{a_1}{\Gamma(\gamma_1+1)} \int_1^{\eta_1} \left(\log \frac{\eta_1}{s}\right)^{\gamma_1} s^{-(\lambda+1)} ds, \tag{7}$$

$$B_1 = \frac{a_2}{\Gamma(\gamma_2+1)} \int_1^{\eta_2} \left(\log \frac{\eta_2}{s}\right)^{\gamma_2} s^{-(\lambda+1)} ds, \quad B_2 = \frac{b_2}{\lambda}(1 - T^{-\lambda}). \tag{8}$$

Proof. In view of Theorem 1 and lemma 2, the general solution of the system (2) can be written as

$$u(t) = c_0 t^{-\lambda} + \frac{c_1}{\lambda}(1 - t^{-\lambda}) + t^{-\lambda} \int_1^t s^{\lambda-1} \mathcal{I}^{\alpha-1} h_1(s) ds, \tag{9}$$

$$v(t) = d_0 t^{-\lambda} + \frac{d_1}{\lambda}(1 - t^{-\lambda}) + t^{-\lambda} \int_1^t s^{\lambda-1} \mathcal{I}^{\beta-1} h_2(s) ds, \tag{10}$$

where $c_i, d_i (i = 0, 1)$ are unknown arbitrary constants. Using the data $u(1) = 0$, $v(1) = 0$ given by (3) in (9) and (10), we find that $c_0 = 0$ and $d_0 = 0$. Thus (9) and (10) take the form:

$$u(t) = \frac{c_1}{\lambda}(1 - t^{-\lambda}) + t^{-\lambda} \int_1^t s^{\lambda-1} \mathcal{I}^{\alpha-1} h_1(s) ds, \tag{11}$$

$$v(t) = \frac{d_1}{\lambda}(1 - t^{-\lambda}) + t^{-\lambda} \int_1^t s^{\lambda-1} \mathcal{I}^{\beta-1} h_2(s) ds. \tag{12}$$

Using the nonlocal integral boundary conditions: $a_1 \mathcal{I}^{\gamma_1} v(\eta_1) + b_1 u(T) = K_1$ and $a_2 \mathcal{I}^{\gamma_2} u(\eta_2) + b_2 v(T) = K_2$ in (11) and (12), we obtain

$$A_1 c_1 + A_2 d_1 = \mathcal{J}_1, \quad B_1 c_1 + B_2 d_1 = \mathcal{J}_2, \tag{13}$$

where A_i and B_i $(i = 1, 2)$ are respectively given by (7) and (8), and

$$\mathcal{J}_1 = K_1 - \frac{a_1}{\Gamma(\gamma_1)} \int_1^{\eta_1} \left(\log \frac{\eta_1}{s}\right)^{\gamma_1-1} s^{-(\lambda+1)} \left(\int_1^s m^{\lambda-1} \mathcal{I}^{\beta-1} h_2(m) dm\right) ds$$

$$- b_1 T^{-\lambda} \int_1^T s^{\lambda-1} \mathcal{I}^{\alpha-1} h_1(s) ds, \tag{14}$$

$$\mathcal{J}_2 = K_2 - \frac{a_2}{\Gamma(\gamma_2)} \int_1^{\eta_2} \left(\log \frac{\eta_2}{s}\right)^{\gamma_2-1} s^{-(\lambda+1)} \left(\int_1^s m^{\lambda-1} \mathcal{I}^{\alpha-1} h_1(m) dm\right) ds$$

$$- b_2 T^{-\lambda} \int_1^T s^{\lambda-1} \mathcal{I}^{\beta-1} h_2(s) ds. \tag{15}$$

Solving the system (13) for c_1 and d_1, we find that

$$c_1 = \frac{(K_2 A_2 - K_1 B_2)}{\Delta} + \frac{T^{-\lambda}}{\Delta}\left[b_1 B_2 \int_1^T s^{\lambda-1} \mathcal{I}^{\alpha-1} h_1(s)ds - b_2 A_2 \int_1^T s^{\lambda-1} \mathcal{I}^{\beta-1} h_2(s)ds\right]$$
$$+ \frac{a_1 B_2}{\Delta \Gamma(\gamma_1)} \int_1^{\eta_1} \left(\log \frac{\eta_1}{s}\right)^{\gamma_1 - 1} s^{-(\lambda+1)} \left(\int_1^s m^{\lambda-1} \mathcal{I}^{\beta-1} h_2(m) dm\right) ds \qquad (16)$$
$$- \frac{a_2 A_2}{\Delta \Gamma(\gamma_2)} \int_1^{\eta_2} \left(\log \frac{\eta_2}{s}\right)^{\gamma_2 - 1} s^{-(\lambda+1)} \left(\int_1^s m^{\lambda-1} \mathcal{I}^{\alpha-1} h_1(m) dm\right) ds,$$

$$d_1 = \frac{(K_1 B_1 - K_2 A_1)}{\Delta} + \frac{T^{-\lambda}}{\Delta}\left[b_2 A_1 \int_1^T s^{\lambda-1} \mathcal{I}^{\beta-1} h_2(s)ds - b_1 B_1 \int_1^T s^{\lambda-1} \mathcal{I}^{\alpha-1} h_1(s)ds\right]$$
$$+ \frac{a_2 A_1}{\Delta \Gamma(\gamma_2)} \int_1^{\eta_2} \left(\log \frac{\eta_2}{s}\right)^{\gamma_2 - 1} s^{-(\lambda+1)} \left(\int_1^s m^{\lambda-1} \mathcal{I}^{\alpha-1} h_1(m) dm\right) ds \qquad (17)$$
$$- \frac{a_1 B_1}{\Delta \Gamma(\gamma_1)} \int_1^{\eta_1} \left(\log \frac{\eta_1}{s}\right)^{\gamma_1 - 1} s^{-(\lambda+1)} \left(\int_1^s m^{\lambda-1} \mathcal{I}^{\beta-1} h_2(m) dm\right) ds,$$

where Δ is given by (6). Substituting the values of c_1 and d_1 in (11) and (12), we obtain the solution (4) and (5). This completes the proof. □

3. Existence and Uniqueness Results

This section is concerned with the main results of the paper. First of all, we fix our terminology. Let $X = \{x : x \in C([1,T], \mathbb{R})$ and $^C\mathcal{D}^\xi x \in C([1,T], \mathbb{R})\}$ and $Y = \{y : y \in C([1,T], \mathbb{R})$ and $^C\mathcal{D}^\xi y \in C([1,T], \mathbb{R})\}$ be the spaces respectively equipped with the norms $\|x\|_X = \|x\| + \|^C\mathcal{D}^\xi x\| = \sup_{t\in[1,T]} |x(t)| + \sup_{t\in[1,T]} |^C\mathcal{D}^\xi x(t)|$ and $\|y\|_Y = \|y\| + \|^C\mathcal{D}^\xi y\| = \sup_{t\in[1,T]} |y(t)| + \sup_{t\in[1,T]} |^C\mathcal{D}^\xi y(t)|$. Observe that $(X, \|.\|_X)$ and $(Y, \|.\|_Y)$ are Banach spaces. In consequence, the product space $(X \times Y, \|.\|_{X\times Y})$ is a Banach space endowed with the norm $\|(x,y)\|_{X\times Y} = \|x\|_X + \|y\|_Y$ for $(x,y) \in X \times Y$.

Using Lemma 3, we introduce an operator $T : X \times Y \to X \times Y$ as follows:

$$T(u,v)(t) := (T_1(u,v)(t), T_2(u,v)(t)), \qquad (18)$$

where

$$T_1(u,v)(t) = \frac{(1-t^{-\lambda})}{\lambda \Delta}\left\{(K_2 A_2 - K_1 B_2) + T^{-\lambda}\left[b_1 B_2 \int_1^T s^{\lambda-1} \mathcal{I}^{\alpha-1} f(s, u(s), v(s), {}^C\mathcal{D}^\xi v(s)) ds\right.\right.$$
$$\left. - b_2 A_2 \int_1^T s^{\lambda-1} \mathcal{I}^{\beta-1} g(s, u(s), {}^C\mathcal{D}^\xi u(s), v(s)) ds\right]$$
$$+ \frac{a_1 B_2}{\Gamma(\gamma_1)} \int_1^{\eta_1} \left(\log \frac{\eta_1}{s}\right)^{\gamma_1 - 1} s^{-(\lambda+1)} \left(\int_1^s m^{\lambda-1} \mathcal{I}^{\beta-1} g(m, u(m), {}^C\mathcal{D}^\xi u(m), v(m)) dm\right) ds$$
$$\left. - \frac{a_2 A_2}{\Gamma(\gamma_2)} \int_1^{\eta_2} \left(\log \frac{\eta_2}{s}\right)^{\gamma_2 - 1} s^{-(\lambda+1)} \left(\int_1^s m^{\lambda-1} \mathcal{I}^{\alpha-1} f(m, u(m), v(m), {}^C\mathcal{D}^\xi v(m)) dm\right) ds\right\}$$
$$+ t^{-\lambda} \int_1^t s^{\lambda-1} \mathcal{I}^{\alpha-1} f(s, u(s), v(s), {}^C\mathcal{D}^\xi v(s)) ds, \qquad (19)$$

$$T_2(u,v)(t) = \frac{(1-t^{-\lambda})}{\lambda \Delta}\left\{(K_1 B_1 - K_2 A_1) + T^{-\lambda}\left[b_2 A_1 \int_1^T s^{\lambda-1} \mathcal{I}^{\beta-1} g(s, u(s), {}^C\mathcal{D}^\xi u(s), v(s)) ds\right.\right.$$
$$\left. - b_1 B_1 \int_1^T s^{\lambda-1} \mathcal{I}^{\alpha-1} f(s, u(s), v(s), {}^C\mathcal{D}^\xi v(s)) ds\right]$$
$$+ \frac{a_2 A_1}{\Gamma(\gamma_2)} \int_1^{\eta_2} \left(\log \frac{\eta_2}{s}\right)^{\gamma_2 - 1} s^{-(\lambda+1)} \left(\int_1^s m^{\lambda-1} \mathcal{I}^{\alpha-1} f(m, u(m), v(m), {}^C\mathcal{D}^\xi v(m)) dm\right) ds \qquad (20)$$
$$\left. - \frac{a_1 B_1}{\Gamma(\gamma_1)} \int_1^{\eta_1} \left(\log \frac{\eta_1}{s}\right)^{\gamma_1 - 1} s^{-(\lambda+1)} \left(\int_1^s m^{\lambda-1} \mathcal{I}^{\beta-1} g(m, u(m), {}^C\mathcal{D}^\xi u(m), v(m)) dm\right) ds\right\}$$

$$+ \ t^{-\lambda} \int_1^t s^{\lambda-1} \mathcal{I}^{\beta-1} g(s, u(s), {}^C\mathcal{D}^{\tilde{\xi}} u(s), v(s)) ds.$$

Next we enlist the assumptions that we need in the sequel.

(H_1) Let $f, g : [1, T] \times \mathbb{R}^3 \to \mathbb{R}$ be continuous functions and there exist real constants $\mu_j, \tau_j \geq 0$ ($j = 1, 2, 3$) and $\mu_0 > 0$, $\tau_0 > 0$ such that

$$|f(t, x_1, x_2, x_3)| \leq \mu_0 + \mu_1|x_1| + \mu_2|x_2| + \mu_3|x_3|,$$
$$|g(t, x_1, x_2, x_3)| \leq \tau_0 + \tau_1|x_1| + \tau_2|x_2| + \tau_3|x_3|, \forall x_j \in \mathbb{R}, j = 1, 2, 3.$$

(H_2) There exist positive constants l, l_1 such that

$$|f(t, x_1, x_2, x_3) - f(t, y_1, y_2, y_3)| \leq l(|x_1 - y_1| + |x_2 - y_2| + |x_3 - y_3|),$$
$$|g(t, x_1, x_2, x_3) - g(t, y_1, y_2, y_3)| \leq l_1(|x_1 - y_1| + |x_2 - y_2| + |x_3 - y_3|), \forall t \in [1, T], x_j, y_j \in \mathbb{R}.$$

For computational convenience, we set

$$\rho = \sup_{t \in [1,T]} \left|1 - t^{-\lambda}\right| = \left|1 - T^{-\lambda}\right|, \tag{21}$$

$$\Theta_1 = \frac{\rho|K_2 A_2 - K_1 B_2|}{\lambda|\Delta|}, \quad \overline{\Theta}_1 = \frac{|K_2 A_2 - K_1 B_2|}{|\Delta|} (\log T)^{1-\tilde{\xi}}, \tag{22}$$

$$\Theta_2 = \frac{\rho|K_1 B_1 - K_2 A_1|}{\lambda|\Delta|}, \quad \overline{\Theta}_2 = \frac{|K_1 B_1 - K_2 A_1|}{|\Delta|} (\log T)^{1-\tilde{\xi}}, \tag{23}$$

$$M_1 = \frac{\rho}{\lambda|\Delta|\Gamma(\alpha+1)} \left[|b_1||B_2|(\log T)^\alpha + \frac{|a_2||A_2|}{\Gamma(\gamma_2+1)}(\log \eta_2)^{\alpha+\gamma_2}\right] + \frac{(\log T)^\alpha}{\Gamma(\alpha+1)}, \tag{24}$$

$$\overline{M}_1 = \frac{(\log T)^{1-\tilde{\xi}}}{|\Delta|\Gamma(\alpha+1)} \left[|b_1||B_2|(\log T)^\alpha + \frac{|a_2||A_2|}{\Gamma(\gamma_2+1)}(\log \eta_2)^{\alpha+\gamma_2} + \lambda|\Delta|(\log T)^\alpha + \alpha|\Delta|(\log T)^{\alpha-1}\right], \tag{25}$$

$$M_2 = \frac{\rho}{\lambda|\Delta|\Gamma(\beta+1)} \left[\frac{|a_1||B_2|}{\Gamma(\gamma_1+1)}(\log \eta_1)^{\beta+\gamma_1} + |b_2||A_2|(\log T)^\beta\right], \tag{26}$$

$$\overline{M}_2 = \frac{(\log T)^{1-\tilde{\xi}}}{|\Delta|\Gamma(\beta+1)} \left[|b_2||A_2|(\log T)^\beta + \frac{|a_1||B_2|}{\Gamma(\gamma_1+1)}(\log \eta_1)^{\beta+\gamma_1}\right], \tag{27}$$

$$N_1 = \frac{\rho}{\lambda|\Delta|\Gamma(\alpha+1)} \left[|b_1||B_1|(\log T)^\alpha + \frac{|a_2||A_1|}{\Gamma(\gamma_2+1)}(\log \eta_2)^{\alpha+\gamma_2}\right], \tag{28}$$

$$\overline{N}_1 = \frac{(\log T)^{1-\tilde{\xi}}}{|\Delta|\Gamma(\alpha+1)} \left[|b_1||B_1|(\log T)^\alpha + \frac{|a_2||A_1|}{\Gamma(\gamma_2+1)}(\log \eta_2)^{\alpha+\gamma_2}\right], \tag{29}$$

$$N_2 = \frac{\rho}{\lambda|\Delta|\Gamma(\beta+1)} \left[\frac{|a_1||B_1|}{\Gamma(\gamma_1+1)}(\log \eta_1)^{\beta+\gamma_1} + |b_2||A_1|(\log T)^\beta\right] + \frac{(\log T)^\beta}{\Gamma(\beta+1)}, \tag{30}$$

$$\overline{N}_2 = \frac{(\log T)^{1-\tilde{\xi}}}{|\Delta|\Gamma(\beta+1)} \left[|b_2||A_1|(\log T)^\beta + \frac{|a_1||B_1|}{\Gamma(\gamma_1+1)}(\log \eta_1)^{\beta+\gamma_1} + \lambda|\Delta|(\log T)^\beta + \beta|\Delta|(\log T)^{\beta-1}\right], \tag{31}$$

$$\varpi_1 = \Theta_1 + \Theta_2 + \frac{\overline{\Theta}_1}{\Gamma(2-\tilde{\xi})} + \frac{\overline{\Theta}_2}{\Gamma(2-\tilde{\xi})} + \mu_0\left(M_1 + N_1 + \frac{\overline{M}_1}{\Gamma(2-\tilde{\xi})} + \frac{\overline{N}_1}{\Gamma(2-\tilde{\xi})}\right)$$
$$+ \tau_0\left(M_2 + N_2 + \frac{\overline{M}_2}{\Gamma(2-\tilde{\xi})} + \frac{\overline{N}_2}{\Gamma(2-\tilde{\xi})}\right), \tag{32}$$

$$\varpi_2 = \mu_1\left(M_1 + N_1 + \frac{\overline{M}_1}{\Gamma(2-\tilde{\xi})} + \frac{\overline{N}_1}{\Gamma(2-\tilde{\xi})}\right) + \max\{\tau_1, \tau_2\}\left(M_2 + N_2 + \frac{\overline{M}_2}{\Gamma(2-\tilde{\xi})} + \frac{\overline{N}_2}{\Gamma(2-\tilde{\xi})}\right), \tag{33}$$

$$\varpi_3 = \max\{\mu_2, \mu_3\}\left(M_1 + N_1 + \frac{\overline{M}_1}{\Gamma(2-\tilde{\xi})} + \frac{\overline{N}_1}{\Gamma(2-\tilde{\xi})}\right) + \tau_3\left(M_2 + N_2 + \frac{\overline{M}_2}{\Gamma(2-\tilde{\xi})} + \frac{\overline{N}_2}{\Gamma(2-\tilde{\xi})}\right). \tag{34}$$

Now, we are in a position to present our first existence result for the boundary value problem (1), which is based on Leray-Schauder alternative.

Lemma 4 (Leray-Schauder alternative [30]). *Let $F : E \to E$ be a completely continuous operator. Let $\varepsilon(F) = \{x \in E : x = \kappa F(x) \text{ for some } 0 < \kappa < 1\}$. Then either the set $\varepsilon(F)$ is unbounded or F has at least one fixed point.*

Theorem 2. *Assume that (H_1) holds and that $\max\{\varpi_2, \varpi_3\} < 1$, where ϖ_2 and ϖ_3 are given by (33) and (34) respectively. Then the boundary value problem (1) has at least one solution on $[1, T]$.*

Proof. In the first step, we establish that the operator $T : X \times Y \to X \times Y$ is completely continuous. By continuity of the functions f and g, it follows that the operators T_1 and T_2 are continuous. In consequence, the operator T is continuous. In order to show that the operator T is uniformly bounded, let $\Omega \subset X \times Y$ be a bounded set. Then there exist positive constants L_1 and L_2 such that $|f(t, u(t), v(t), {}^C\mathcal{D}^{\bar{\xi}} v(t))| \leq L_1$, $|g(t, u(t), {}^C\mathcal{D}^{\bar{\xi}} u(t), v(t))| \leq L_2$, $\forall (u, v) \in \Omega$. Then, for any $(u, v) \in \Omega$, we have

$$\begin{aligned}
|T_1(u,v)(t)| &\leq \frac{|K_2 A_2 - K_1 B_2|\rho}{\lambda |\Delta|} + \frac{\rho L_1}{\lambda |\Delta|} \bigg\{ \frac{|b_1||B_2|T^{-\lambda}}{\Gamma(\alpha-1)} \int_1^T s^{\lambda-1} \bigg(\int_1^s \Big(\log \frac{s}{m} \Big)^{\alpha-2} \frac{dm}{m} \bigg) ds \\
&+ \frac{|a_2||A_2|}{\Gamma(\gamma_2)\Gamma(\alpha-1)} \int_1^{\eta_2} \Big(\log \frac{\eta_2}{s} \Big)^{\gamma_2-1} s^{-(\lambda+1)} \bigg(\int_1^s m^{\lambda-1} \Big(\log \frac{m}{r} \Big)^{\alpha-2} \frac{dr}{r} \bigg) dm \bigg) ds \bigg\} \\
&+ \frac{L_1 |t^{-\lambda}|}{\Gamma(\alpha-1)} \int_1^t s^{\lambda-1} \bigg(\int_1^s \Big(\log \frac{s}{m} \Big)^{\alpha-2} \frac{dm}{m} \bigg) ds \\
&+ \frac{\rho L_2}{\lambda |\Delta|} \bigg\{ \frac{|b_2||A_2|T^{-\lambda}}{\Gamma(\beta-1)} \int_1^T s^{\lambda-1} \bigg(\int_1^s \Big(\log \frac{s}{m} \Big)^{\beta-2} \frac{dm}{m} \bigg) ds \\
&+ \frac{|a_1||B_2|}{\Gamma(\gamma_1)\Gamma(\beta-1)} \int_1^{\eta_1} \Big(\log \frac{\eta_1}{s} \Big)^{\gamma_1-1} s^{-(\lambda+1)} \bigg(\int_1^s m^{\lambda-1} \Big(\log \frac{m}{r} \Big)^{\beta-2} \frac{dr}{r} \bigg) dm \bigg) ds \bigg\}, \\
&\leq \frac{|K_2 A_2 - K_1 B_2|\rho}{\lambda |\Delta|} + \frac{\rho L_1}{\lambda |\Delta| \Gamma(\alpha+1)} \Big[|b_1||B_2|(\log T)^{\alpha} + \frac{|a_2||A_2|}{\Gamma(\gamma_2+1)} \big(\log \eta_2 \big)^{\alpha+\gamma_2} \Big] \\
&+ \frac{L_1}{\Gamma(\alpha+1)} (\log T)^{\alpha} + \frac{\rho L_2}{\lambda |\Delta| \Gamma(\beta+1)} \Big[\frac{|a_1||B_2|}{\Gamma(\gamma_1+1)} \big(\log \eta_1 \big)^{\beta+\gamma_1} + |b_2 A_2| (\log T)^{\beta} \Big],
\end{aligned}$$

which, on taking the norm for $t \in [1, T]$ and using (22), (24) and (26) yields

$$\|T_1(u,v)\| \leq \Theta_1 + L_1 M_1 + L_2 M_2.$$

Since $0 < \bar{\xi} < 1$, we use Remark 1 to get

$$|{}^C\mathcal{D}^{\bar{\xi}} T_1(u,v)(t)| \leq \frac{1}{\Gamma(1-\bar{\xi})} \int_1^t \Big(\log \frac{t}{s} \Big)^{-\bar{\xi}} |T_1'(u,v)(s)| \frac{ds}{s} \leq \frac{1}{\Gamma(2-\bar{\xi})} \left(\overline{\Theta}_1 + L_1 \overline{M}_1 + L_2 \overline{M}_2 \right),$$

where $\overline{\Theta}_1$, $\overline{M}_1$ and $\overline{M}_2$ are respectively given by (22), (25) and (27). Hence

$$\|T_1(u,v)\|_X = \|T_1(u,v)\| + \|{}^C\mathcal{D}^{\bar{\xi}} T_1(u,v)\| \leq \Theta_1 + L_1 M_1 + L_2 M_2 + \frac{1}{\Gamma(2-\bar{\xi})} \left(\overline{\Theta}_1 + L_1 \overline{M}_1 + L_2 \overline{M}_2 \right). \quad (35)$$

Similarly, using (23), (28) and (30), we obtain

$$|T_2(u,v)(t)| \leq \frac{\rho |K_1 B_1 - K_2 A_1|}{\lambda |\Delta|} + \frac{\rho L_1}{\lambda |\Delta| \Gamma(\alpha+1)} \Big[|b_1||B_1|(\log T)^{\alpha} + \frac{|a_2||A_1|}{\Gamma(\gamma_2+1)} \big(\log \eta_2 \big)^{\alpha+\gamma_2} \Big]$$

$$+ \frac{\rho L_2}{\lambda |\Delta| \Gamma(\beta+1)} \left[\frac{|b_2||A_1|}{\Gamma(\gamma_1+1)} \left(\log \eta_1 \right)^{\beta+\gamma_1} + |b_2||A_1| \left(\log T \right)^{\beta} \right] + \frac{L_2}{\Gamma(\beta+1)} \left(\log T \right)^{\beta}$$
$$\leq \Theta_2 + L_1 N_1 + L_2 N_2.$$

As before, one can find that

$$|{}^C\mathcal{D}^{\tilde{\xi}} T_2(u,v)(t)| \leq \frac{1}{\Gamma(2-\tilde{\xi})} \left(\overline{\Theta}_2 + L_1 \overline{N}_1 + L_2 \overline{N}_2 \right),$$

where $\overline{\Theta}_2$, $\overline{N}_1$ and $\overline{N}_2$ are respectively given by (23), (29) and (31).

In consequence, we get

$$\|T_2(u,v)\|_Y = \|T_2(u,v)\| + \|{}^C\mathcal{D}^{\tilde{\xi}} T_2(u,v)\| \leq \Theta_2 + L_1 N_1 + L_2 N_2 + \frac{1}{\Gamma(2-\tilde{\xi})} \left(\overline{\Theta}_2 + L_1 \overline{N}_1 + L_2 \overline{N}_2 \right). \quad (36)$$

From the inequalities (35) and (36), we deduce that T_1 and T_2 are uniformly bounded, which implies that the operator T is uniformly bounded.

Next, we show that T is equicontinuous. Let $t_1, t_2 \in [1,T]$ with $t_1 < t_2$. Then we have

$$|T_1(u,v)(t_2) - T_1(u,v)(t_1)|$$
$$\leq \frac{\left|t_1^{-\lambda} - t_2^{-\lambda}\right|}{\lambda |\Delta|} \Bigg\{ |K_2 A_2 - K_1 B_2| + T^{-\lambda} \bigg[|b_1||B_2| \int_1^T s^{\lambda-1} \mathcal{I}^{\alpha-1} |f(s, u(s), v(s), {}^C\mathcal{D}^{\tilde{\xi}} v(s))| ds$$
$$+ |b_2||A_2| \int_1^T s^{\lambda-1} \mathcal{I}^{\beta-1} |g(s, u(s), {}^C\mathcal{D}^{\tilde{\xi}} u(s), v(s))| ds \bigg]$$
$$+ \frac{|a_1||B_2|}{\Gamma(\gamma_1)} \int_1^{\eta_1} \left(\log \frac{\eta_1}{s} \right)^{\gamma_1 - 1} s^{-(\lambda+1)} \left(\int_1^s m^{\lambda-1} \mathcal{I}^{\beta-1} |g(m, u(m), {}^C\mathcal{D}^{\tilde{\xi}} u(m), v(m))| dm \right) ds$$
$$+ \frac{|a_2||A_2|}{\Gamma(\gamma_2)} \int_1^{\eta_2} \left(\log \frac{\eta_2}{s} \right)^{\gamma_2 - 1} s^{-(\lambda+1)} \left(\int_1^s m^{\lambda-1} \mathcal{I}^{\alpha-1} |f(m, u(m), v(m), {}^C\mathcal{D}^{\tilde{\xi}} v(m))| dm \right) ds \Bigg\}$$
$$+ \left|t_2^{-\lambda} - t_1^{-\lambda}\right| \int_1^{t_1} s^{\lambda-1} \mathcal{I}^{\alpha-1} |f(s, u(s), v(s), {}^C\mathcal{D}^{\tilde{\xi}} v(s))| ds$$
$$+ t_2^{-\lambda} \int_{t_1}^{t_2} s^{\lambda-1} \mathcal{I}^{\alpha-1} |f(s, u(s), v(s), {}^C\mathcal{D}^{\tilde{\xi}} v(s))| ds$$
$$\to 0 \text{ as } t_2 \to t_1,$$

independent of (u,v) on account of $|f(t, u(t), v(t), {}^C\mathcal{D}^{\tilde{\xi}} v(t))| \leq L_1$ and $|g(t, u(t), {}^C\mathcal{D}^{\tilde{\xi}} u(t), v(t))| \leq L_2$. Also we have

$$|{}^C\mathcal{D}^{\tilde{\xi}} T_1(u,v)(t_2) - {}^C\mathcal{D}^{\tilde{\xi}} T_1(u,v)(t_1)|$$
$$\leq \frac{1}{\Gamma(2-\tilde{\xi})} \left| \int_1^{t_2} \left(\log \frac{t_2}{s} \right)^{-\tilde{\xi}} T_1'(u,v)(s) ds - \int_1^{t_1} \left(\log \frac{t_1}{s} \right)^{-\tilde{\xi}} T_1'(u,v)(s) ds \right|$$
$$\leq \frac{1}{\Gamma(1-\tilde{\xi})} \Bigg\{ \int_1^{t_1} \left| \left(\log \frac{t_2}{s} \right)^{-\tilde{\xi}} - \left(\log \frac{t_1}{s} \right)^{-\tilde{\xi}} \right| s^{-\lambda-1} ds + \int_{t_1}^{t_2} \left(\log \frac{t_2}{s} \right)^{-\tilde{\xi}} s^{-(\lambda+1)} ds \Bigg\} \times$$
$$\times \Bigg\{ |K_2 A_2 - K_1 B_2| + T^{-\lambda} \bigg[|b_1||B_2| \int_1^T s^{\lambda-1} \mathcal{I}^{\alpha-1} |f(s, u(s), v(s), {}^C\mathcal{D}^{\tilde{\xi}} v(s))| ds$$
$$+ |b_2||A_2| \int_1^T s^{\lambda-1} \mathcal{I}^{\beta-1} |g(s, u(s), {}^C\mathcal{D}^{\tilde{\xi}} u(s), v(s))| ds \bigg]$$
$$+ \frac{|a_1||B_2|}{\Gamma(\gamma_1)} \int_1^{\eta_1} \left(\log \frac{\eta_1}{s} \right)^{\gamma_1 - 1} s^{-(\lambda+1)} \left(\int_1^s m^{\lambda-1} \mathcal{I}^{\beta-1} |g(m, u(m), {}^C\mathcal{D}^{\tilde{\xi}} u(m), v(m))| dm \right) ds$$

$$+ \frac{|a_2||A_2|}{\Gamma(\gamma_2)} \int_1^{\eta_2} \left(\log \frac{\eta_2}{s}\right)^{\gamma_2-1} s^{-(\lambda+1)} \left(\int_1^s m^{\lambda-1} \mathcal{I}^{\alpha-1} |f(m,u(m),v(m),{}^C\mathcal{D}^{\bar{\xi}}v(m))| dm\right) ds \bigg\}$$

$$+ \frac{\lambda}{\Gamma(1-\bar{\xi})} \int_1^{t_1} \left|\left(\log \frac{t_2}{s}\right)^{-\bar{\xi}}\right.$$

$$- \left(\log \frac{t_1}{s}\right)^{-\bar{\xi}}\bigg| s^{-(\lambda+1)} \left(\int_1^s m^{\lambda-1} \mathcal{I}^{\alpha-1} |f(m,u(m),v(m),{}^C\mathcal{D}^{\bar{\xi}}v(m))| dm\right) ds$$

$$+ \frac{\lambda}{\Gamma(1-\bar{\xi})} \int_{t_1}^{t_2} \left(\log \frac{t_2}{s}\right)^{-\bar{\xi}} s^{-(\lambda+1)} \left(\int_1^s m^{\lambda-1} \mathcal{I}^{\alpha-1} |f(m,u(m),v(m),{}^C\mathcal{D}^{\bar{\xi}}v(m))| dm\right) ds$$

$$+ \frac{1}{\Gamma(1-\bar{\xi})} \int_1^{t_1} \left|\left(\log \frac{t_2}{s}\right)^{-\bar{\xi}} - \left(\log \frac{t_1}{s}\right)^{-\bar{\xi}}\right| s^{-1} \mathcal{I}^{\alpha-1} |f(s,u(s),v(s),{}^C\mathcal{D}^{\bar{\xi}}v(s))| ds$$

$$+ \frac{1}{\Gamma(1-\bar{\xi})} \int_{t_1}^{t_2} \left(\log \frac{t_2}{s}\right)^{-\bar{\xi}} s^{-1} \mathcal{I}^{\alpha-1} |f(s,u(s),v(s),{}^C\mathcal{D}^{\bar{\xi}}v(s))| ds \to 0 \text{ as } t_2 \to t_1,$$

independent of (u,v). In a similar manner, one can obtain that

$$|T_2(u,v)(t_2) - T_2(u,v)(t_1)| \to 0 \text{ and } |{}^C\mathcal{D}^{\xi}T_2(u,v)(t_2) - {}^C\mathcal{D}^{\xi}T_2(u,v)(t_1)| \to 0$$

as $t_2 \to t_1$ independent of (u,v) on account of the boundedness of f and g. Thus the operator T is equicontinuous in view of equicontinuity of T_1 and T_2. Therefore, by Arzela-Ascoli's theorem, it follows that the operator T is compact (completely continuous).

Finally, it will be shown that the set $\varepsilon(T) = \{(u,v) \in X \times Y : (u,v) = \kappa T(u,v) ; 0 \le \kappa \le 1\}$ is bounded. Let $(u,v) \in \varepsilon(T)$. Then $(u,v) = \kappa T(u,v)$. For any $t \in [1,T]$, we have $u(t) = \kappa T_1(u,v)(t)$, $v(t) = \kappa T_2(u,v)(t)$. Using (H_1) in (19), we get

$$|u(t)|$$
$$\le \frac{\rho}{\lambda|\Delta|} \bigg\{ |K_2 A_2 - K_1 B_2| + T^{-\lambda} \bigg[\frac{|b_1||B_2|}{\Gamma(\alpha-1)} \int_1^T s^{\lambda-1} \bigg(\int_1^s \left(\log \frac{s}{m}\right)^{\alpha-2} \times$$

$$\times \bigg(\mu_0 + \mu_1|u(m)| + \mu_2|v(m)| + \mu_3|{}^C\mathcal{D}^{\bar{\xi}}v(m)|\bigg) \frac{dm}{m}\bigg) ds$$

$$+ \frac{|b_2||A_2|}{\Gamma(\beta-1)} \int_1^T s^{\lambda-1} \bigg(\int_1^s \left(\log \frac{s}{m}\right)^{\beta-2} \bigg(\tau_0 + \tau_1|u(m)| + \tau_2|{}^C\mathcal{D}^{\xi}u(m)| + \tau_3|v(m)|\bigg) \frac{dm}{m}\bigg) ds\bigg]$$

$$+ \frac{|a_1||B_2|}{\Gamma(\gamma_1)\Gamma(\beta-1)} \int_1^{\eta_1} \left(\log \frac{\eta_1}{s}\right)^{\gamma_1-1} s^{-(\lambda+1)} \times$$

$$\times \bigg(\int_1^s m^{\lambda-1} \bigg(\int_1^m \left(\log \frac{m}{r}\right)^{\beta-2} \big[\tau_0 + \tau_1|u(r)| + \tau_2|{}^C\mathcal{D}^{\xi}u(r)| + \tau_3|v(r)|\big] \frac{dr}{r}\bigg) dm\bigg) ds$$

$$+ \frac{|a_2||A_2|}{\Gamma(\gamma_2)\Gamma(\alpha-1)} \int_1^{\eta_2} \left(\log \frac{\eta_2}{s}\right)^{\gamma_2-1} s^{-(\lambda+1)} \bigg(\int_1^s m^{\lambda-1} \bigg(\int_1^m \left(\log \frac{m}{r}\right)^{\alpha-2} \times$$

$$\times \bigg(\mu_0 + \mu_1|u(r)| + \mu_2|v(r)| + \mu_3|{}^C\mathcal{D}^{\bar{\xi}}v(r)|\bigg) \frac{dr}{r}\bigg) dm\bigg) ds\bigg\}$$

$$+ \frac{|t^{-\lambda}|}{\Gamma(\alpha-1)} \int_1^t s^{\lambda-1} \bigg(\int_1^s \left(\log \frac{s}{m}\right)^{\alpha-2} \big[\mu_0 + \mu_1|u(m)| + \mu_2|v(m)| + \mu_3|{}^C\mathcal{D}^{\bar{\xi}}v(m)|\big] \frac{dm}{m}\bigg) ds,$$

which, on taking the norm for $t \in [1,T]$, yields

$$\|u\| \le \Theta_1 + \Big(\mu_0 + \mu_1\|u\|_X + \max\{\mu_2,\mu_3\}\|v\|_Y\Big) M_1$$
$$+ \Big(\tau_0 + \max\{\tau_1,\tau_2\}\|u\|_X + \tau_3\|v\|_Y\Big) M_2.$$

Similarly one can find that

$$\|{}^C\mathcal{D}^{\bar{\xi}}u\| \le \frac{1}{\Gamma(2-\bar{\xi})}\Big\{\overline{\Theta}_1 + \Big(\mu_0 + \mu_1\|u\|_X + \max\{\mu_2,\mu_3\}\|v\|_Y\Big)\overline{M}_1 \\ + \Big(\tau_0 + \max\{\tau_1,\tau_2\}\|u\|_X + \tau_3\|v\|_Y\Big)\overline{M}_2\Big\}.$$

Consequently, we have

$$\begin{aligned}\|u\|_X &= \|u\| + \|{}^C\mathcal{D}^{\bar{\xi}}u\| \\ &\le \Theta_1 + \frac{\overline{\Theta}_1}{\Gamma(2-\bar{\xi})} + \Big(M_1 + \frac{\overline{M}_1}{\Gamma(2-\bar{\xi})}\Big)\Big(\mu_0 + \mu_1\|u\|_X + \max\{\mu_2,\mu_3\}\|v\|_Y\Big) \\ &+ \Big(M_2 + \frac{\overline{M}_2}{\Gamma(2-\bar{\xi})}\Big)\Big(\tau_0 + \max\{\tau_1,\tau_2\}\|u\|_X + \tau_3\|v\|_Y\Big).\end{aligned} \quad (37)$$

Likewise, we can derive that

$$\begin{aligned}\|v\|_Y &\le \Theta_2 + \frac{\overline{\Theta}_2}{\Gamma(2-\bar{\xi})} + \Big(N_1 + \frac{\overline{N}_1}{\Gamma(1-\bar{\xi})}\Big)\Big(\mu_0 + \mu_1\|u\|_X + \max\{\mu_2,\mu_3\}\|v\|_Y\Big) \\ &+ \Big(N_2 + \frac{\overline{N}_2}{\Gamma(2-\bar{\xi})}\Big)\Big(\tau_0 + \max\{\tau_1,\tau_2\}\|u\|_X + \tau_3\|v\|_Y\Big).\end{aligned} \quad (38)$$

From (37) and (38), we get

$$\begin{aligned}\|u\|_X + \|v\|_Y &= \Theta_1 + \Theta_2 + \frac{\overline{\Theta}_1}{\Gamma(2-\bar{\xi})} + \frac{\overline{\Theta}_2}{\Gamma(2-\bar{\xi})} \\ &+ \mu_0\Big(M_1 + N_1 + \frac{\overline{M}_1}{\Gamma(2-\bar{\xi})} + \frac{\overline{N}_1}{\Gamma(2-\bar{\xi})}\Big) + \tau_0\Big(M_2 + N_2 + \frac{\overline{M}_2}{\Gamma(2-\bar{\xi})} + \frac{\overline{N}_2}{\Gamma(1-\bar{\xi})}\Big) \\ &+ \|u\|_X\Big[\mu_1\Big(M_1 + N_1 + \frac{\overline{M}_1}{\Gamma(2-\bar{\xi})} + \frac{\overline{N}_1}{\Gamma(2-\bar{\xi})}\Big) + \max\{\tau_1,\tau_2\}\Big(M_2 + N_2 + \frac{\overline{M}_2}{\Gamma(2-\bar{\xi})} + \frac{\overline{N}_2}{\Gamma(2-\bar{\xi})}\Big)\Big] \\ &+ \|v\|_X\Big[\max\{\mu_2,\mu_3\}\Big(M_1 + N_1 + \frac{\overline{M}_1}{\Gamma(2-\bar{\xi})} + \frac{\overline{N}_1}{\Gamma(2-\bar{\xi})}\Big) + \tau_3\Big(M_2 + N_2 + \frac{\overline{M}_2}{\Gamma(2-\bar{\xi})} + \frac{\overline{N}_2}{\Gamma(2-\bar{\xi})}\Big)\Big] \\ &\le \varpi_1 + \max\{\varpi_2,\varpi_3\}\|(u,v)\|_{X\times Y},\end{aligned} \quad (39)$$

which, together with $\|(u,v)\|_{X\times Y} = \|u\|_X + \|v\|_Y$, yields

$$\|(u,v)\|_{X\times Y} \le \frac{\varpi_1}{1-\max\{\varpi_2,\varpi_3\}}.$$

This shows that $\varepsilon(T)$ is bounded. Thus, Lemma 4 applies and that T has at least one fixed point. This implies that the boundary value problem (1) has at least one solution on $[1,T]$. The proof is completed. □

Example 1. *Consider the following coupled system of Caputo-Hadamard type sequential fractional differential equations*

$$\begin{aligned}({}^C\mathcal{D}^{\frac{3}{2}} + \frac{1}{2}{}^C\mathcal{D}^{\frac{1}{2}})x(t) &= f(t,x(t),y(t),{}^C\mathcal{D}^{\frac{1}{3}}y(t)), \quad t\in[1,10], \\ ({}^C\mathcal{D}^{\frac{5}{4}} + \frac{1}{2}{}^C\mathcal{D}^{\frac{1}{4}})y(t) &= g(t,x(t),{}^C\mathcal{D}^{\frac{1}{4}}x(t),y(t)), \quad t\in[1,10],\end{aligned} \quad (40)$$

equipped with nonlocal coupled non-conserved boundary conditions:

$$\begin{aligned}u(1) &= 0, \ -2\mathcal{I}^{\frac{3}{2}}v(2) + u(10) = 3, \\ v(1) &= 0, \ -\mathcal{I}^{\frac{1}{4}}u(3) + 2v(10) = 7.\end{aligned} \quad (41)$$

Here, $\lambda = 1/2, \alpha = 3/2, \beta = 5/4, T = 10, a_1 = -2, a_2 = -1, b_1 = 1, b_2 = 2, K_1 = 3, K_2 = 7, \eta_1 = 2, \eta_2 = 3, \gamma_1 = 3/2, \gamma_2 = 1/4, \xi = 1/3, \bar{\xi} = 1/4,$

$$f(t, x(t), y(t), {}^{C}\mathcal{D}^{\frac{1}{3}}y(t)) = \frac{1}{2(24+t^2)}\left(3(t-1) + \frac{1}{2}\sin(x(t)) + |y(t)| + |{}^{C}\mathcal{D}^{\frac{1}{3}}y(t))|\right)$$

and

$$g(t, x(t), {}^{C}\mathcal{D}^{\frac{1}{4}}x(t), y(t)) = \frac{1}{49t}\left(\frac{1-t}{2} + |x(t)| + \frac{|{}^{C}\mathcal{D}^{\frac{1}{4}}x(t)|}{1+|{}^{C}\mathcal{D}^{\frac{1}{4}}x(t)|} + \sin(y(t))\right).$$

Clearly, the functions f and g satisfy the condition (H_1) with $\mu_0 = \frac{27}{50}, \mu_1 = \frac{1}{100}, \mu_2 = \mu_3 = \frac{1}{50}, \tau_0 = \frac{9}{98}, \tau_1 = \tau_2 = \tau_3 = \frac{1}{49}$. Using the given data, we find that $A_1 \approx 1.3675, |A_2| \approx 0.2186, |B_1| \approx 0.7865, B_2 \approx 2.7351, |\Delta| \approx 3.5684, \rho \approx 0.6838, \Theta_1 \approx 2.5581, \overline{\Theta}_1 \approx 3.49653, \Theta_2 \approx 2.76436, \overline{\Theta}_2 \approx 3.52477, M_1 \approx 5.4654, M_2 \approx 0.9275, \overline{M}_1 \approx 9.5348, \overline{M}_2 \approx 1.2677, N_1 \approx 1.8178, N_2 \approx 5.2756, \overline{N}_1 \approx 1.6640, \overline{N}_2 \approx 7.9915, \omega_1 \approx 25.0711, \omega_2 \approx 0.530375, \omega_3 \approx 0.725385$. With $\max\{\omega_2, \omega_3\} < 1$, all the conditions of Theorem 2 are satisfied. Therefore, the problem (40) and (41) has a solution on on $[1, 10]$.

The next result deals with the uniqueness of solutions for the problem (1) and relies on Banach contraction mapping principle. For computational convenience, we introduce the notations:

$$\begin{aligned}
\Phi_1 &= \Theta_1 + r_1 M_1 + r_2 M_2, \ \Psi_1 = \ell M_1 + \ell_1 M_2, \ \Phi_2 = \Theta_2 + r_1 N_1 + r_2 N_2, \ \Psi_2 = \ell N_1 + \ell_1 N_2,\\
\overline{\Phi}_1 &= \overline{\Theta}_1 + r_1 \overline{M}_1 + r_2 \overline{M}_2, \ \overline{\Psi}_1 = \ell \overline{M}_1 + \ell_1 \overline{M}_2, \ \overline{\Phi}_2 = \overline{\Theta}_2 + r_1 \overline{N}_1 + r_2 \overline{N}_2, \ \overline{\Psi}_2 = \ell \overline{N}_1 + \ell_1 \overline{N}_2,\\
r_1 &= \sup_{t \in [1,T]} f(t, 0, 0, 0) < \infty, \ r_2 = \sup_{t \in [1,T]} g(t, 0, 0, 0) < \infty.
\end{aligned} \quad (42)$$

Theorem 3. *Assume that (H_2) holds. Then the boundary value problem (1) has a unique solution on $[1, T]$, provided that*

$$\Psi_1 + \frac{\overline{\Psi}_1}{\Gamma(2-\bar{\xi})} < \frac{1}{2} \text{ and } \Psi_2 + \frac{\overline{\Psi}_2}{\Gamma(2-\xi)} < \frac{1}{2}, \quad (43)$$

where Ψ_i and $\overline{\Psi}_i$ $(i = 1, 2)$ are given by (42).

Proof. Let us fix

$$r \geq \max\left\{\frac{\Phi_1 + \frac{\overline{\Phi}_1}{\Gamma(2-\bar{\xi})}}{\frac{1}{2} - (\Psi_1 + \frac{\overline{\Psi}_1}{\Gamma(2-\bar{\xi})})}, \frac{\Phi_2 + \frac{\overline{\Phi}_2}{\Gamma(2-\xi)}}{\frac{1}{2} - (\Psi_2 + \frac{\overline{\Psi}_2}{\Gamma(2-\xi)})}\right\},$$

where $\Phi_i, \overline{\Phi}_i$, and $\Psi_i, \overline{\Psi}_i$ $(i = 1, 2)$ are given by (42). Then we show that $TB_r \subset B_r$, where

$$B_r = \{(u, v) \in X \times X : \|(u, v)\|_{X \times Y} \leq r\}.$$

For $(u, v) \in B_r$, we have

$$\begin{aligned}
|f(t, u(t), v(t), {}^{C}\mathcal{D}^{\xi}v(t))| &\leq |f(t, u(t), v(t), {}^{C}\mathcal{D}^{\xi}v(t)) - f(t, 0, 0, 0)| + |f(t, 0, 0, 0)|\\
&\leq \ell[|u(t)| + |v(t)| + |{}^{C}\mathcal{D}^{\xi}v(t)|] + r_1\\
&\leq \ell[\|u\|_X + \|v\|_Y] + r_1 \leq \ell\|(u, v)\|_{X \times Y} + r_1 \leq \ell r + r_1.
\end{aligned}$$

Similarly, we can find that

$$|g(t, u(t), {}^{C}\mathcal{D}^{\bar{\xi}}u(t), v(t))| \leq \ell_1 r + r_2.$$

Then

$$|T_1(u, v)(t)| \leq \Theta_1 + r_1 M_1 + r_2 M_2 + (\ell M_1 + \ell_1 M_2) r \leq \Phi_1 + \Psi_1 r,$$

and

$$|{}^C\mathcal{D}^{\tilde{\xi}}T_1(u,v)(t)| \leq \frac{1}{\Gamma(2-\tilde{\xi})}\left[\overline{\Theta}_1 + r_1\overline{M}_1 + r_2\overline{M}_2 + (\ell\overline{M}_1 + \ell_1\overline{M}_2)r\right] \leq \frac{1}{\Gamma(2-\tilde{\xi})}\left[\overline{\Phi}_1 + \overline{\Psi}_1 r\right].$$

Therefore,

$$\|T_1(u,v)\|_X = \|T_1(u,v)\| + \|{}^C\mathcal{D}^{\tilde{\xi}}T_1(u,v)\| \leq \Phi_1 + \frac{\overline{\Phi}_1}{\Gamma(2-\tilde{\xi})} + \left[\Psi_1 + \frac{\overline{\Psi}_1}{\Gamma(2-\tilde{\xi})}\right]r \leq \frac{r}{2}. \qquad (44)$$

In similar manner, we obtain

$$|T_2(u,v)(t)| \leq \Phi_2 + \Psi_2 r, \qquad |{}^C\mathcal{D}^{\tilde{\xi}}T_2(u,v)(t)| \leq \frac{1}{\Gamma(2-\tilde{\xi})}\left[\overline{\Phi}_2 + \overline{\Psi}_2 r\right].$$

In consequence, we get

$$\|T_2(u,v)\|_Y = \|T_2(u,v)\| + \|{}^C\mathcal{D}^{\tilde{\xi}}T_2(u,v)\| \leq \Phi_2 + \frac{\overline{\Phi}_2}{\Gamma(2-\tilde{\xi})} + \left[\Psi_2 + \frac{\overline{\Psi}_2}{\Gamma(2-\tilde{\xi})}\right]r \leq \frac{r}{2}. \qquad (45)$$

Thus, it follows from (44) and (45) that

$$\|T(u,v)\|_{X\times Y} = \|T_1(u,v)\|_X + \|T_2(u,v)\|_X \leq r,$$

which implies that $TB_r \subset B_r$.

Next we show that the operator T is a contraction. For that, let $u_i, v_i \in B_r$ ($i=1,2$). Then, for each $t \in [1,T]$, we have

$$
\begin{aligned}
&|T_1(u_1,v_1)(t) - T_1(u_2,v_2)(t)| \\
&\leq \frac{|1-t^{-\lambda}|}{\lambda|\Delta|}\Bigg\{T^{-\lambda}\Bigg[\frac{|b_1|B_2|}{\Gamma(\alpha-1)}\int_1^T s^{\lambda-1}\left(\int_1^s \left(\log\frac{s}{m}\right)^{\alpha-2}\times\right. \\
&\quad \times \left|f(m,u_1(m),v_1(m),{}^C\mathcal{D}^{\tilde{\xi}}v_1(m)) - f(m,u_2(m),v_2(m),{}^C\mathcal{D}^{\tilde{\xi}}v_2(m))\right|\frac{dm}{m}\bigg)ds \\
&\quad + \frac{|b_2A_2|}{\Gamma(\beta-1)}\int_1^T s^{\lambda-1}\left(\int_1^s\left(\log\frac{s}{m}\right)^{\beta-2}\times\right. \\
&\quad \times \left|g(m,u_1(m),{}^C\mathcal{D}^{\tilde{\xi}}u_1(m),v_1(m)) - g(m,u_2(m),{}^C\mathcal{D}^{\tilde{\xi}}u_2(m),v_2(m))\right|\frac{dm}{m}\bigg)ds\Bigg] \\
&\quad + \frac{|a_1B_2|}{\Gamma(\gamma_1)\Gamma(\beta-1)}\int_1^{\eta_1}\left(\log\frac{\eta_1}{s}\right)^{\gamma_1-1}s^{-(\lambda+1)}\left(\int_1^s m^{\lambda-1}\left(\int_1^m\left(\log\frac{m}{r}\right)^{\beta-2}\times\right.\right. \\
&\quad \times \left|g(r,u_1(r),{}^C\mathcal{D}^{\tilde{\xi}}u_1(r),v_1(r)) - g(r,u_2(r),{}^C\mathcal{D}^{\tilde{\xi}}u_2(r),v_2(r))\right|\frac{dr}{r}\bigg)dm\bigg)ds \\
&\quad + \frac{|a_2A_2|}{\Gamma(\gamma_2)\Gamma(\alpha-1)}\int_1^{\eta_2}\left(\log\frac{\eta_2}{s}\right)^{\gamma_2-1}s^{-(\lambda+1)}\left(\int_1^s m^{\lambda-1}\left(\int_1^m\left(\log\frac{m}{r}\right)^{\alpha-2}\times\right.\right. \\
&\quad \times \left|f(r,u_1(r),v_1(r),{}^C\mathcal{D}^{\tilde{\xi}}v_1(r)) - f(r,u_2(r),v_2(r),{}^C\mathcal{D}^{\tilde{\xi}}v_2(r))\right|\frac{dr}{r}\bigg)dm\bigg)ds\Bigg\} \\
&\quad + \frac{t^{-\lambda}}{\Gamma(\alpha-1)}\int_1^t s^{\lambda-1}\left(\int_1^s\left(\log\frac{s}{m}\right)^{\alpha-2}\times\right. \\
&\quad \times \left|f(m,u_1(m),v_1(m),{}^C\mathcal{D}^{\tilde{\xi}}v_1(m)) - f(m,u_2(m),v_2(m),{}^C\mathcal{D}^{\tilde{\xi}}v_2(m))\right|\frac{dm}{m}\bigg)ds \\
&\leq M_1\ell\left[\|u_1-u_2\| + \|v_1-v_2\| + \|{}^C\mathcal{D}^{\tilde{\xi}}v_1 - {}^C\mathcal{D}^{\tilde{\xi}}v_2\|\right] \\
&\quad + M_2\ell_1\left[\|u_1-u_2\| + \|{}^C\mathcal{D}^{\tilde{\xi}}u_1 - {}^C\mathcal{D}^{\tilde{\xi}}u_2\| + \|v_1-v_2\|\right]
\end{aligned}
$$

$$\leq \quad \Psi_1\left[\|u_1 - u_2\|_X + \|v_1 - v_2\|_Y\right].$$

Also we have

$$|{}^C\mathcal{D}^{\bar{\xi}} T_1(u_1,v_1)(t) - {}^C\mathcal{D}^{\bar{\xi}} T_1(u_2,v_2)(t)| \leq \frac{1}{\Gamma(1-\bar{\xi})} \int_1^t \left(\log \frac{t}{s}\right)^{-\bar{\xi}} |T_1'(u_1,v_1)(s) - T_1'(u_2,v_2)(s)| ds$$

$$\leq \frac{\Psi_1}{\Gamma(2-\bar{\xi})}[\|u_1 - u_2\|_X + \|v_1 - v_2\|_Y].$$

From the foregoing inequalities, we get

$$\|T_1(u_1,v_1) - T_1(u_2,v_2)\|_X = \|T_1(u_1,v_1) - T_1(u_2,v_2)\| + \|{}^C\mathcal{D}^{\bar{\xi}} T_1(u_1,v_1) - {}^C\mathcal{D}^{\bar{\xi}} T_1(u_2,v_2)\|$$

$$\leq \left[\Psi_1 + \frac{\Psi_1}{\Gamma(1-\bar{\xi})}\right][\|u_1 - u_2\|_X + \|v_1 - v_2\|_Y]. \tag{46}$$

Similarly, we can find that

$$\|T_2(u_1,v_1) - T_2(u_2,v_2)\|_Y \leq \left[\Psi_2 + \frac{\Psi_2}{\Gamma(2-\bar{\zeta})}\right][\|u_1 - u_2\|_X + \|v_1 - v_2\|_Y] \tag{47}$$

Consequently, it follows from (46) and (47) that

$$\|T(u_1,v_1) - T(u_2,v_2)\|_{X\times Y} = \|T_1(u_1,v_1) - T_1(u_2,v_2)\|_X + \|T_2(u_1,v_1) - T_2(u_2,v_2)\|_X$$

$$\leq \left[\Psi_1 + \Psi_2 + \frac{\Psi_1}{\Gamma(2-\bar{\xi})} + \frac{\Psi_2}{\Gamma(2-\bar{\zeta})}\right][\|u_1 - u_2\|_X + \|v_1 - v_2\|_Y].$$

This shows that T is a contraction by (43). Hence, by Banach fixed point theorem, the operator T has a unique fixed point which corresponds to a unique solution of problem (1). This completes the proof. □

Example 2. *Consider the following coupled system of fractional differential equations*

$$({}^C\mathcal{D}^{\frac{3}{2}} + \frac{1}{2}{}^C\mathcal{D}^{\frac{1}{2}})x(t) = \frac{1}{2(24+t^2)}\left(3 + \sin(x(t)) + |y(t)| + \tan^{-1}({}^C\mathcal{D}^{\frac{1}{3}}y(t))\right), \quad t \in [1,10]$$

$$({}^C\mathcal{D}^{\frac{5}{4}} + \frac{1}{2}{}^C\mathcal{D}^{\frac{1}{4}})y(t) = \frac{1}{49t}\left(\frac{t}{2} + |x(t)| + \frac{|{}^C\mathcal{D}^{\frac{1}{4}}x(t)|}{1+|{}^C\mathcal{D}^{\frac{1}{4}}x(t)|} + \sin(y(t))\right), \tag{48}$$

supplemented with nonlocal coupled non-conserved boundary conditions:

$$\begin{aligned} u(1) &= 0, \quad -2\mathcal{I}^{\frac{3}{2}}v(2) + u(10) = 3, \\ v(1) &= 0, \quad -\mathcal{I}^{\frac{1}{4}}u(3) + 2v(10) = 7. \end{aligned} \tag{49}$$

Here, $\lambda = 1/2, \alpha = 3/2, \beta = 5/4, T = 10, a_1 = -2, a_2 = -1, b_1 = 1, b_2 = 2, K_1 = 3, K_2 = 7, \eta_1 = 2, \eta_2 = 3, \gamma_1 = 3/2, \gamma_2 = 1/4, \xi = 1/3, \bar{\zeta} = 1/4$,

$$f(t,x(t),y(t),{}^C\mathcal{D}^{\bar{\xi}}y(t)) = \frac{1}{2(24+t^2)}(3+\sin(x(t))+|y(t)|+\tan^{-1}({}^C\mathcal{D}^{\frac{1}{3}}y(t)))$$

and

$$g(t,x(t),{}^C\mathcal{D}^{\bar{\zeta}}x(t),y(t)) = \frac{1}{49t}\left(\frac{t}{2} + |x(t)| + \frac{|{}^C\mathcal{D}^{\frac{1}{4}}x(t)|}{1+|{}^C\mathcal{D}^{\frac{1}{4}}x(t)|} + \sin(y(t))\right).$$

From the inequalities:

$$|f(t,x_1(t),y_1(t),{}^C\mathcal{D}^{\frac{1}{3}}y_1(t)) - f(t,x_2(t),y_2(t),{}^C\mathcal{D}^{\frac{1}{3}}y_2(t))|$$

$$\leq \frac{1}{50}\left(|x_1(t)-x_2(t)|+|y_1(t)-y_2(t)|+|{}^C\mathcal{D}^{\frac{1}{3}}y_1(t)-{}^C\mathcal{D}^{\frac{1}{3}}y_2(t)|\right),$$

$$|g(t,x_1(t),{}^C\mathcal{D}^{\frac{1}{4}}x_1(t),y_1(t))-g(t,x_2(t),{}^C\mathcal{D}^{\frac{1}{4}}x_2(t),y_2(t))|$$
$$\leq \frac{1}{49}\left(|x_1(t)-x_2(t)|+|{}^C\mathcal{D}^{\frac{1}{4}}x_1(t)-{}^C\mathcal{D}^{\frac{1}{4}}x_2(t)|+|y_1(t)-y_2(t)|\right),$$

we have $l=\frac{1}{50}$ and $l_1=\frac{1}{49}$. Using the given data, we find that $A_1\approx 1.3675, |A_2|\approx 0.2186$, $|B_1|\approx 0.7865$, $B_2\approx 2.7351, |\Delta|\approx 3.5684, \rho\approx 0.6838, M_1\approx 5.4654, M_2\approx 0.9275, \Psi_1\approx 0.1282, \overline{M}_1\approx 9.5348, \overline{M}_2\approx 1.2677, \overline{\Psi}_1\approx 0.2166, N_1\approx 1.8178, N_2\approx 5.2756, \Psi_2\approx 0.1439, \overline{N}_1\approx 1.6640, \overline{N}_2\approx 7.9915, \overline{\Psi}_2\approx 0.1964.$ Further

$$\Psi_1+\frac{\overline{\Psi}_1}{\Gamma(7/4)}\approx 0.3639 < 0.5,\quad \Psi_2+\frac{\overline{\Psi}_2}{\Gamma(5/3)}\approx 0.3615 < 0.5.$$

Thus all the conditions of Theorem 3 are satisfied. In consequence, by Theorem 3, there exists a unique solution for the problem (48) and (49) on $[1,10]$.

4. Conclusions

We have developed the existence theory for a nonlocal integral boundary value problem of coupled sequential fractional differential equations involving Caputo-Hadamard fractional derivatives and Hadamard fractional integrals. Several results follow as special cases by fixing the values of the parameters involved in the problem. For example, by taking $a_i=-1, b_i=1, K_1=0=K_2$ and $T=e$, our results correspond to the ones associated with coupled strip boundary conditions of the form:

$$u(1)=0,\quad u(T)=\mathcal{I}^{\gamma_1}v(\eta_1),\ \gamma_1>0,\ 1<\eta_1<e,$$
$$v(1)=0,\quad v(T)=\mathcal{I}^{\gamma_2}u(\eta_2),\ \gamma_2>0,\ 1<\eta_2<e.$$

If we take $a_1=0=a_2$ in the results of this paper, we obtain the ones for a coupled system of Caputo-Hadamard fractional differential equations and uncoupled Dirichlet boundary conditions. We emphasize that the main results as well as the special cases presented in this paper are new and enrich the existing literature on the topic.

Author Contributions: Conceptualization, B.A.; data curation, S.A.; formal analysis, S.A., B.A. and A.A.; methodology, S.A., B.A. and A.A. All authors have read and agreed to the published version of the manuscript.

Funding: This research received no external funding.

Acknowledgments: We thank the reviewers for their positive remarks on our work.

Conflicts of Interest: The authors declare no conflict of interest.

References

1. Di Paola, M.; Pinnola, F.P.; Zingales, M. Fractional differential equations and related exact mechanical models. *Comput. Math. Appl.* **2013**, *66*, 608–620. [CrossRef]
2. Ahmed, N.; Vieru, D.; Fetecau, C.; Shah, N.A. Convective flows of generalized time-nonlocal nanofluids through a vertical rectangular channel. *Phys. Fluids* **2018**, *30*, 052002. [CrossRef]
3. Tarasov, V.E. *Fractional Dynamics: Applications of Fractional Calculus to Dynamics of Particles, Fields and Media*; Springer: New York, NY, USA, 2010.
4. Tarasova, V.V.; Tarasov, V.E. Logistic map with memory from economic model. *Chaos Solitons Fractals* **2017**, *95*, 84–91. [CrossRef]
5. Hadamard, J. Essai sur l'etude des fonctions donnees par leur developpment de Taylor. *J. Math. Pure Appl. Ser.* **1892**, *8*, 101–186.
6. Ahmad, B.; Alsaedi, A.; Ntouyas, S.K.; Tariboon, J. *Hadamard-Type Fractional Differential Equations, Inclusions and Inequalities*; Springer International Publishing: Cham, Switzerland, 2017.

7. Kilbas, A.A.; Srivastava, H.M.; Trujillo, J.J. *Theory and Applications of Fractional Differential Equations*; North-Holland Mathematics Studies, 204; Elsevier Science B.V.: Amsterdam, The Netherlands, 2006.
8. Podlubny, I. *Fractional Differential Equations*; Academic Press: San Diego, CA, USA, 1999.
9. Samko, S.G.; Kilbas, A.A.; Marichev, O.I. *Fractional Integrals and Derivatives: Theory and Applications*; Gordon and Breach: Yverdon, Switzerland, 1993.
10. Ahmad, B.; Ntouyas, S.K. A fully Hadamard type integral boundary value problem of a coupled system of fractional differential equations. *Fract. Calc. Appl. Anal.* **2014**, *17*, 348–360. [CrossRef]
11. Ahmad, B.; Ntouyas, S.K.; Alsaedi A. Sequential fractional differential equations and inclusions with semi-periodic and nonlocal integro-multipoint boundary conditions. *J. King Saud Univ. Sci.* **2019**, *31*, 184–193. [CrossRef]
12. Aljoudi, S.; Ahmad, B.; Nieto, J.J.; Alsaedi, A. On coupled Hadamard type sequential fractional differential equations with variable coefficients and nonlocal integral boundary conditions. *Filomat* **2017**, *31*, 6041–6049. [CrossRef]
13. Aljoudi, S.; Ahmad, B.; Nieto, J.J.; Alsaedi, A. A coupled system of Hadamard type sequential fractional differential equations with coupled strip conditions. *Chaos Solitons Fractals* **2016**, *91*, 39–46. [CrossRef]
14. Garra, R.; Polito, F. On some operators involving Hadamard derivatives. *Integral Transforms Spec. Funct.* **2013**, *24*, 773–782. [CrossRef]
15. Jiang, J.; Liu, L. Existence of solutions for a sequential fractional differential system with coupled boundary conditions. *Bound. Value Probl.* **2016**. [CrossRef]
16. Jiang, J.; O'Regan, D.; Xu, J.; Fu, Z. Positive solutions for a system of nonlinear Hadamard fractional differential equations involving coupled integral boundary conditions. *J. Inequal. Appl.* **2019**. [CrossRef]
17. Kilbas, A.A. Hadamard-type fractional calculus. *J. Korean Math. Soc.* **2001**, *38*, 1191–1204.
18. Ma, Q.; Wang, R.; Wang, J.; Ma, Y. Qualitative analysis for solutions of a certain more generalized two-dimensional fractional differential system with Hadamard derivative. *Appl. Math. Comput.* **2015**, *257*, 436–445. [CrossRef]
19. Wang, J.R.; Zhou, Y.; Medved, M. Existence and stability of fractional differential equations with Hadamard derivative. *Topol. Methods Nonlinear Anal.* **2013**, *41*, 113–133.
20. Wang, J.R.; Zhang, Y. On the concept and existence of solutions for fractional impulsive systems with Hadamard derivatives. *Appl. Math. Lett.* **2015**, *39*, 85–90. [CrossRef]
21. Jarad, J.; Abdeljawad, T.; Baleanu, D. Caputo-type modification of the Hadamard fractional derivatives. *Adv. Differ. Equ.* **2012**. [CrossRef]
22. Abbas, S.; Benchohra, M.; Hamidi, N.; Henderson, J. Caputo-Hadamard fractional differential equations in Banach spaces. *Fract. Calc. Appl. Anal.* **2018**, *21*, 1027–1045. [CrossRef]
23. Ahmad, B.; Ntouyas, S.K. Existence and uniqueness of solutions for Caputo-Hadamard sequential fractional order neutral functional differential equations. *Electron. J. Differ. Equ.* **2017**, *36*, 1–11.
24. Benchohra, M.; Bouriah, S.; Graef, J.R. Boundary Value Problems for Non-linear Implicit Caputo-Hadamard-Type Fractional Differential Equations with Impulses. *Mediterr. J. Math.* **2017**, *14*, 206–216. [CrossRef]
25. Cichon, M.; Salem, H.A.H. On the solutions of Caputo-Hadamard Pettis-type fractional differential equations. *Rev. R. Acad. Cienc. Exactas Fís. Nat. Ser. A Mat. RACSAM* **2019**, *113*, 3031–3053. [CrossRef]
26. Wang, G.; Ren, X.; Zhang, L.; Ahmad, B. Explicit iteration and unique positive solution for a Caputo-Hadamard fractional turbulent flow model. *IEEE Access* **2019**, *7*, 109833–109839. [CrossRef]
27. Yukunthorn, W.; Ahmad, B.; Ntouyas, S.K.; Tariboon, J. On Caputo-Hadamard type fractional impulsive hybrid systems with nonlinear fractional integral conditions. *Nonlinear Anal. Hybrid Syst.* **2016**, *19*, 77–92. [CrossRef]
28. Zhang, X. On impulsive partial differential equations with Caputo-Hadamard fractional derivatives. *Adv. Differ. Equ.* **2016**. [CrossRef]
29. Bai, Y.; Kong, H. Existence of solutions for nonlinear Caputo-Hadamard fractional differential equations via the method of upper and lower solutions. *J. Nonlinear Sci. Appl.* **2017**, *10*, 5744–5752. [CrossRef]
30. Granas, A.; Dugundji, J. *Fixed Point Theory*; Springer: New York, NY, USA, 2003.

 © 2020 by the authors. Licensee MDPI, Basel, Switzerland. This article is an open access article distributed under the terms and conditions of the Creative Commons Attribution (CC BY) license (http://creativecommons.org/licenses/by/4.0/).

Article

Multi-Strip and Multi-Point Boundary Conditions for Fractional Langevin Equation

Ahmed Salem * and Balqees Alghamdi

Department of Mathematics, Faculty of Science, King Abdulaziz University, P.O.Box 80203, Jeddah 21589, Saudi Arabia; aisharif@ju.edu.sa
* Correspondence: asaalshreef@kau.edu.sa

Received: 9 March 2020; Accepted: 24 April 2020; Published: 28 April 2020

Abstract: In the present paper, we discuss a new boundary value problem for the nonlinear Langevin equation involving two distinct fractional derivative orders with multi-point and multi-nonlocal integral conditions. The fixed point theorems for Schauder and Krasnoselskii–Zabreiko are applied to study the existence results. The uniqueness of the solution is given by implementing the Banach fixed point theorem. Some examples showing our basic results are provided.

Keywords: fixed point theorem; fractional Langevin equations; existence and uniqueness solution

MSC: 26A33; 34A08; 34A12; 34B15

1. Introduction

The Langevin equation was discovered by Langevin a century ago to render an accurate description of the evolution of physical phenomena in fluctuating environments. This equation can be considered as a special form of the generalized Langevin equation [1], which has turned into a modern research project theme.

Fractional calculus has attracted many authors and researchers in many different scientific disciplines. Many of the recent advances in fractional calculus were motivated by the modern applications of fractional integro-differential equations in various fields, in particular physics. One of the main reasons for its popularity in modeling various transport properties in complex heterogeneous and disordered media is that it provides a natural setting for describing processes with memory and is fractal or multi-fractal in nature [2]. For systems in complex media, the ordinary Langevin equation does not provide the correct description of the dynamics. Various generalizations of Langevin equations have been proposed to describe dynamical processes in a fractal medium. One such generalization is the Langevin equation with two fractional orders, which incorporates the fractal and memory properties with a dissipative memory kernel into the Langevin equation. This possible extension requires the replacement of the ordinary derivative by a fractional derivative in the Langevin equation to give the fractional Langevin equation [3–5]. Various versions of fractional Langevin-type equations have been proposed to model anomalous diffusion [6,7], and both deterministic and stochastic fractional equations are used to describe non-Debye dielectric relaxation phenomena.

Anomalous diffusion has been found in various physical and biological systems. The mean squared displacement of the particle shows a power law dependence on time $\langle x^2(t) \rangle \sim t^\alpha$, becoming subdiffusion in the case $0 < \alpha < 1$, superdiffusion for $\alpha > 1$, and normal classical diffusion for $\alpha = 1$ [8–10]. Several stochastic approaches to anomalous diffusion exist. In most cases, such behavior is considered to be connected with the self-similar properties of the diffusion medium. As this took place, the generalized Langevin equation came to fame following Kubos' work and the related fractional Brownian motion, originally introduced by Kolmogorov [11] and popularized by Mandelbrot [12]. It is remarkable to note that Mainardi and Pironi [13] introduced a fractional Langevin equation as a particular case of a generalized Langevin equation and for the first time represented the velocity and displacement correlation functions in terms of the Mittag–Leffler functions.

Due to the extremely useful role of the fractional Langevin equation in applied mathematics, physics, engineering, and several branches of science, it has acquired many scientific contributions in the field of finding exact solutions [3,14,15], approximate solutions through numerical analysis methods [16–18], and studying the existence and uniqueness of the solution (see [19–25] and the references given therein). Studying differential equations with integral boundary conditions designates an extremely useful and interesting class of boundary value problems. Several of the problems are in chemical engineering, population dynamics, heat conduction, thermoelasticity, underground water flow, and plasma physics [26]. Furthermore, there are many published contributions concerned with the fractional boundary value problem with the integral boundary conditions (see [27–29] and the references given therein). Multi-point boundary value problems for differential equations become apparent naturally in scientific applications. For an illustration, given a dynamical system with m degrees of freedom, there may be available exactly m cases spotted at m distinct times. A mathematical depiction of such problems is in an m-point boundary value problem. Multi-point problems for differential equations are a special class of interface problems, and hence solvable with various techniques. Studying fractional differential equations with multi-pint boundary condition has been drawn the attention of many contributors (see [30–33] and the references given therein).

Motivated by the significance of the integral boundary conditions and fractional Langevin equations in different branches of science and engineering, this paper is interested in studying the nonlinear fractional Langevin equation:

$$^cD^p(^cD^q + \mu)x(t) = f(t, x(t), {}^cD^r x(t)), \qquad t \in [0,1] \tag{1}$$

with the new auxiliary multi-integral and multi-point boundary conditions:

$$x(0) = 0, \quad D^q x(0) = 0, \quad x(1) = \sum_{i=1}^{n} \alpha_i x(\eta_i) + \sum_{i=1}^{n} \beta_i \int_0^{\eta_i} x(s) ds \tag{2}$$

where $^cD^p, {}^cD^q$, and $^cD^r$ are the Liouville–Caputo fractional derivative of orders $p \in (1,2], q \in (0,1]$, and $0 < r \le q$, $\mu \in \mathbb{R}$ is the dissipative parameter, $\mu, \alpha_i, \beta_i \in \mathbb{R}$, $\eta_i \in (0,1), i = 1,2,\cdots,n$ such that $n \in \mathbb{N}$ with $\omega = \sum_{i=1}^{n} \alpha_i \eta_i^{q+1} \ne 1$, and the function $f : [0,1] \times \mathbb{R} \times \mathbb{R} \to \mathbb{R}$ is continuous.

It is worth mentioning that $x(t)$ in Equation (1) is the displacement of the particle in the general interval $t \in [0, a]$, $a > 0$ (for simplicity, we apply the transformation $t = t/a$ to make t in the unit interval $[0,1]$). Instead of the ordinary definition of the velocity and acceleration as the first and second derivatives of the displacement, respectively, we render fractional forms $^cD^\alpha x(t)$, $0 < \alpha \le 1$ for the velocity and $^cD^\beta x(t)$, $1 < \beta \le 2$ for the the acceleration. The product of two fractional derivatives gives the term $^cD^{\alpha+\beta}$, which represents the acceleration if $1 < \alpha + \beta \le 2$ and the aberrancy of the curve or the Jerk term [34–36] if $2 < \alpha + \beta \le 3$ instead of defining it as a third time derivative of the displacement. We take here the function f in the general form, which is constituted by the position $x(t)$ and velocity $^cD^r x(t)$ of the particle at time t. This function may contain an external force field, a position-dependent phenomenological fluid friction coefficient, the intensity of the stochastic force, or the zero-mean Gaussian white noise term.

The first and second boundary conditions in Equation (2) indicate that the particle begins its motion from stillness at the origin. The last condition in Equation (2), which seems to be a linear combination of the values of the unknown function at the multi-point and multi-strip, can be interpreted as "the value of of the unknown function at the terminal point proportionate to the summation of values of it at midst nonlocal m-points and the areas under its curve from the initial point to the midst points".

In mathematical analysis, the existence and uniqueness of the solution for differential and integral equations have become major topics. There are many fixed point theorems used to discuss the existence results [37]. One of these theorems due to Krasnoselskii–Zabreiko lacks usage, although it gives more precise sufficient conditions of the existence results. This inspired us to discuss the existence of the solution for our problem by implementing the Krasnoselskii–Zabreiko fixed point theorem. Furthermore, we apply the Schauder fixed point under different assumptions and show its applicability by means of studying a numerical example. In addition, the uniqueness of the solution is investigated by applying the Banach fixed point theorem.

The strategy of the paper is as follows: In the section below, we render some definitions and results that are needed in this paper. The existence and uniqueness are discussed in Section 3. In the last section, we establish some examples to show these results.

Introducing the fractional element provides many possibilities for the generalizations of models described in the previous subsection.

2. Preliminaries

Throughout this section, the definitions needed and the notations are given. Let $C[0,1]$ be the class continuous functions on $[a,b]$. Furthermore, let $AC[a,b]$ be the space of functions f that are absolutely continuous on $[a,b]$. For $n \in \mathbb{N}$, we denote by $AC^n[a,b]$ the space of real functions $f(t)$, which have continuous derivatives up to order $n-1$ on $[a,b]$ such that $f^{(n-1)}(t) \in AC[a,b]$. In particular, $AC^1[a,b] = AC[a,b]$ [38].

Definition 1 ([38])**.** *If $x(t) \in C[a,b]$, then, the R-L fractional integral with order $p > 0$ exists almost everywhere on $[a,b]$ and can be represented in the form:*

$$I^p x(t) = \frac{1}{\Gamma(p)} \int_a^t (t-s)^{p-1} x(s) ds.$$

Definition 2 ([38])**.** *If $x(t) \in AC^n[a,b]$ and $n \in \mathbb{N}$, the Liouville–Caputo fractional derivative of order $n-1 < p \leq n$ exists almost everywhere on $[a,b]$ and can be represented in the form:*

$$^c D^p x(t) = \frac{1}{\Gamma(n-p)} \int_a^t (t-s)^{n-p-1} x^{(n)}(s) ds.$$

Lemma 1 ([38,39])**.** *Let $n \in \mathbb{N}$, $n-1 < q \leq n$, and $x(t) \in C^n[0,1]$, then we have:*

$$I^q \, ^c D^q x(t) = x(t) + a_0 + a_1 t + \cdots + a_{n-1} t^{n-1}$$

Lemma 2 ([38,39])**.** *Let $p > 0$, $n \in \mathbb{N}$ such that $n-1 < q \leq n$, then:*

1. $^c D^q I^p x(t) = D^{q-p} x(t)$ if $q > p$,
2. $^c D^q I^p x(t) = I^{p-q} x(t)$ if $p > q$.

Lemma 3. *Suppose the function $g : C[0,1] \to \mathbb{R}$; hence, the unique solution of the linear equation:*

$$^c D^p(^c D^q + \mu) x(t) = g(t), \quad t \in [0,1] \qquad (3)$$

with the conditions mentioned in Equation (2), can be taken the form:

$$x(t) = \frac{1}{\Gamma(q+p)} \int_0^t (t-s)^{q+p-1} g(s) ds - \frac{\mu}{\Gamma(q)} \int_0^t (t-s)^{q-1} x(s) ds$$
$$+ \frac{t^{q+1}}{(1-\omega)} \left[\frac{\mu}{\Gamma(q)} \int_0^1 (1-s)^{q-1} x(s) ds - \frac{1}{\Gamma(p+q)} \int_0^1 (1-s)^{q+p-1} g(s) ds \right.$$
$$+ \sum_{i=1}^n \frac{\alpha_i}{\Gamma(p+q)} \int_0^{\eta_i} (\eta_i - s)^{q+p-1} g(s) ds - \mu \sum_{i=1}^n \frac{\alpha_i}{\Gamma(q)} \int_0^{\eta_i} (\eta_i - s)^{q-1} x(s) ds$$
$$\left. + \sum_{i=1}^n \beta_i \int_0^{\eta_i} x(s) ds \right] \tag{4}$$

Proof. Applying Lemma 1, we get:

$$^c D^q x(t) = I^p g(t) - \mu x(t) + a_0 + a_1 t \tag{5}$$

Furthermore, we apply Lemma 1 and use the relation $I^q t^p = \frac{\Gamma(p+1)}{\Gamma(p+q+1)} t^{p+q}$, and Equation (5) becomes:

$$x(t) = \frac{1}{\Gamma(q+p)} \int_0^t (t-s)^{q+p-1} g(s) ds - \frac{\mu}{\Gamma(q)} \int_0^t (t-s)^{q-1} x(s) ds$$
$$+ \frac{t^q}{\Gamma(q+1)} a_0 + \frac{t^{q+1}}{\Gamma(q+2)} a_1 + a_2 \tag{6}$$

By using the boundary conditions $x(0) = 0$ and $D^q x(0) = 0$ in Equations (5) and (6), respectively, we find that $a_0 = 0$ and $a_2 = 0$. The boundary equation $x(1) = \sum_{i=1}^n \alpha_i x(\eta_i) + \sum_{i=1}^n \beta_i \int_0^{\eta_i} x(s) ds$ in Equation (6) gives the value of the constant a_1 as:

$$a_1 = \frac{\Gamma(q+2)}{(1-\omega)} \left[\mu I^q x(1) - I^{p+q} g(1) + \sum_{i=1}^n \alpha_i I^{p+q} g(\eta_i) - \mu \sum_{i=1}^n \alpha_i I^q x(\eta_i) \right.$$
$$\left. + \sum_{i=1}^n \beta_i \int_0^{\eta_i} x(s) ds \right]$$

Substitute the values a_0, a_1 and a_2 in Equation (6) to obtain Equation (4). Conversely, inserting Equation (4) in the left side of Equation (3) using Lemma 2 implies the right side. Furthermore, it is not difficult to see that Equation (4) verifies the boundary condition Equation (2). This completes the proof. □

3. Main Results

Define the space:

$$X = \{ x : x \in C[0,1], \, ^c D^r x \in C[0,1], 0 < r \leq 1 \}$$

equipped with the norm:

$$\|x\|_X = \|x\| + \|\,^c D^r x(t)\| = \max_{t \in [0,1]} |x(t)| + \max_{t \in [0,1]} |\,^c D^r x(t)|.$$

It is worth pointing out that Su [40] proved that X is a Banach space equipped with the former norm.

Assume the following hypotheses that we need to prove the existence and uniqueness results of the problem Equations (1) and (2).

(G_1) $f : [0,1] \times \mathbb{R} \times \mathbb{R} \to \mathbb{R}$ is a continuous function;

(G_2) There exists a positive function $\psi \in X$ such that
$|f(t,x,y)| \leq \psi(t) + a_1|x|^{r_1} + a_2|y|^{r_2}$ where $a_1, a_2 \in \mathbb{R}^+$ and $0 < r_1, r_2 \leq 1$;

(G_3) The continuous function $f(t,0,0)$ does not vanish identically in $[0,1]$;

(G_4) $\lim_{r \to \infty} \frac{f(t,x(t),y(t))}{x(t)+y(t)} = \kappa(t)$ uniformly in $[0,1]$ where $x, y \in X$, $r = \|x\| + \|y\|$ and $\kappa : [0,1] \to \mathbb{R}$ is continuous;

(G_5) There exists a constant $L > 0$ such that:
$$|f(t,x,y) - f(t,\hat{x},\hat{y})| \leq L(|x - \hat{x}| + |y - \hat{y}|), \quad t \in [0,1], \ x, \hat{x}, y, \hat{y} \in \mathbb{R}.$$

For convenience, let:
$$\Theta = \Theta_{p,r} + \Theta_{p,0}, \tag{7}$$
$$Y = Y_r + Y_0 \tag{8}$$

where:
$$\Theta_{p,r} = \frac{1}{\Gamma(p+q-r+1)} + \frac{\Gamma(q+2)}{|1-\omega|\Gamma(q-r+2)\Gamma(p+q+1)}\left(1 + \sum_{i=1}^{n} \alpha_i \eta_i^{p+q}\right)$$

$$Y_r = |\mu|\Theta_{0,r} + \frac{\Gamma(q+2)}{|1-\omega|\Gamma(q-r+2)} \sum_{i=1}^{n} \beta_i \eta_i$$

We express the operator $T : X \to X$ as:
$$(Tx)(t) = \frac{1}{\Gamma(q+p)} \int_0^t (t-s)^{q+p-1} f(s,x(s), {}^cD^r x(s))ds - \frac{\mu}{\Gamma(q)} \int_0^t (t-s)^{q-1} x(s)ds$$
$$+ \frac{t^{q+1}}{1-\omega}[T_1(x) + T_2(f)] \tag{9}$$

and its rth Caputo fractional derivative:
$${}^cD^r(Tx)(t) = \frac{1}{\Gamma(q+p-r-1)} \int_0^t (t-s)^{q+p-1} f(s,x(s), {}^cD^r x(s))ds$$
$$- \frac{\mu}{\Gamma(q-r)} \int_0^t (t-s)^{q-r-1} x(s)ds + \frac{\Gamma(q+2)t^{q+1-r}}{(1-\omega)\Gamma(q-r+2)}[T_1(x) + T_2(f)] \tag{10}$$

where:
$$T_1(x) = \frac{\mu}{\Gamma(q)} \int_0^1 (1-s)^{q-1} x(s)ds - \mu \sum_{i=1}^{n} \frac{\alpha_i}{\Gamma(q)} \int_0^{\eta_i} (\eta_i - s)^{q-1} x(s)ds + \sum_{i=1}^{n} \beta_i \int_0^{\eta_i} x(s)ds$$

$$T_2(f) = \sum_{i=1}^{n} \frac{\alpha_i}{\Gamma(p+q)} \int_0^{\eta_i} (\eta_i - s)^{q+p-1} f(s,x(s), {}^cD^r x(s))ds$$
$$- \frac{1}{\Gamma(p+q)} \int_0^1 (1-s)^{q+p-1} f(s,x(s), {}^cD^r x(s))ds.$$

It is easy to see that:
$$\|T_1(x)\| \leq \frac{\|x\|}{\Gamma(q+1)}\left(|\mu| + |\mu|\sum_{i=1}^{n}|\alpha_i|\eta_i^q + \Gamma(q+1)\sum_{i=1}^{n}|\beta_i|\eta_i\right) \tag{11}$$

$$\|T_2(f)\| \leq \frac{\|f\|}{\Gamma(p+q+1)}\left(1 + \sum_{i=1}^{n}|\alpha_i|\eta_i^{q+p}\right). \tag{12}$$

25

Our first result is discussing the existence of the solution for the problem by the Schauder fixed point theorem.

Theorem 1. *Assume that G_1 and G_2 hold. Then, the boundary value problem Equations (1) and (2) have a solution.*

Proof. We express the operator $T : X \to X$ and let a closed ball $\mathcal{B}_\xi = \{x \in X : \|x\|_X \leq \xi\}$ taking:

$$\xi > \max\{4\|\psi\|_X \Theta, (4a_1\Theta)^{\frac{1}{1-r_1}}, (4a_2\Theta)^{\frac{1}{1-r_2}}, 4\xi Y\}$$

Then, we claim that $T\mathcal{B}_\xi \subset \mathcal{B}_\xi$. For $x \in \mathcal{B}_\xi$ and by the condition G_1, we give:

$$|(Tx)(t)| \leq \frac{1}{\Gamma(q+p)} \int_0^t (t-s)^{q+p-1} |f(s,x(s),{}^c D^r x(s))| ds + \frac{|\mu|}{\Gamma(q)} \int_0^t (t-s)^{q-1} |x(s)| ds$$
$$+ \frac{1}{|1-\omega|} [|T_1(x)| + |T_2(f)|]$$
$$\leq (\|\psi\|_X + a_1\xi^{r_1} + a_2\xi^{r_2}) \left\{ \frac{t^{p+q}}{\Gamma(p+q+1)} + \frac{1}{(1-\omega)\Gamma(p+q+1)} + \frac{\sum_{i=1}^n |\alpha_i||\eta_i|^{p+q}}{(1-\omega)\Gamma(p+q+1)} \right\}$$
$$+ \|x\|_X \left\{ \frac{|\mu|t^q}{\Gamma(q+1)} + \frac{t^{q+1}|\mu|}{(1-\omega)\Gamma(q+1)} + \frac{t^{q+1}|\mu|}{(1-\omega)\Gamma(q+1)} \sum_{i=1}^n \alpha_i \eta_i^q + \frac{t^{q+1}}{(1-\omega)} \sum_{i=1}^n |\beta_i||\eta_i| \right\}$$
$$\leq (\|\psi\|_X + a_1\xi^{r_1} + a_2\xi^{r_2}) \left\{ \frac{1}{\Gamma(p+q+1)} + \frac{1 + \sum_{i=1}^n |\alpha_i||\eta_i|^{p+q}}{(1-\omega)\Gamma(p+q+1)} \right\}$$
$$+ \xi \left\{ \frac{|\mu|}{\Gamma(q+1)} + \frac{|\mu|(1+\sum_{i=1}^n |\alpha_i||\eta_i^q|)}{(1-\omega)\Gamma(q+1)} + \frac{\sum_{i=1}^n |\beta_i||\eta_i|}{1-\omega} \right\}$$
$$\leq (\|\psi\|_X + a_1\xi^{r_1} + a_2\xi^{r_2})\Theta_{p,0} + \xi Y_0$$

Similarly, we have:

$$|{}^c D^r (Tx)(t)| \leq (\|\psi\|_X + a_1\xi^{r_1} + a_2\xi^{r_2})\Theta_{p,r} + \xi Y_r$$

Consequently,

$$\|Tx\|_X = \max|(Tx)(t)| + \max|({}^c D^r(Tx)(t))|$$
$$\leq (\|\psi\|_X + a_1\xi^{r_1} + a_2\xi^{r_2})(\Theta_{p,0} + \Theta_{p,r}) + \xi(Y_0 + Y_r)$$
$$\leq (\|\psi\|_X + a_1\xi^{r_1} + a_2\xi^{r_2})\Theta + \xi Y$$
$$\leq \frac{\xi}{4} + \frac{\xi}{4} + \frac{\xi}{4} + \frac{\xi}{4} = \xi$$

Then, the operator $T : X \to X$ is uniformly bounded. Next, we show that T is equicontinuous. We set:

$$N = \max_{t \in [0,1]} |f(t, x(t), {}^c D^r x(t))| + 1$$

and for $x \in \mathcal{B}_\xi$, let $t_1, t_2 \in [0,1]$, whereas for $t_1 < t_2$, we get:

$$|(Tx)(t_2) - (Tx)(t_1)| \leq \left| \frac{1}{\Gamma(p+q)} \int_0^{t_2} (t_2-s)^{q+p-1} f(s, x(s), {}^cD^r x(s)) ds \right.$$

$$- \frac{\mu}{\Gamma(q)} \int_0^{t_2} (t_2-s)^{q-1} x(s) ds - \frac{1}{\Gamma(p+q)} \int_0^{t_1} (t_1-s)^{q+p-1} f(s, x(s), D^r x(s)) ds$$

$$+ \frac{\mu}{\Gamma(q)} \int_0^{t_1} (t_1-s)^{q-1} x(s) ds \bigg| + \left| \frac{(t_2^{q+1} - t_1^{q+1})}{1-\omega} (T_1(x) + T_2(f)) \right|$$

$$\leq \left| \frac{1}{\Gamma(p+q)} \int_0^{t_1} [(t_2-s)^{q+p-1} - (t_1-s)^{p+q-1}] f(s, x(s), {}^cD^r x(s)) ds \right.$$

$$+ \frac{1}{\Gamma(p+q)} \int_{t_1}^{t_2} (t_2-s)^{q+p-1} f(s, x(s), {}^cD^r x(s)) ds \bigg|$$

$$+ \left| \frac{\mu}{\Gamma(q)} \int_0^{t_1} [(t_1-s)^{q-1} - (t_2-s)^{q-1}] x(s) ds + \frac{\mu}{\Gamma(q)} \int_{t_1}^{t_2} (t_2-s)^{q-1} x(s) ds \right|$$

$$+ \frac{t_2^{q+1} - t_1^{q+1}}{|1-\omega|} (|T_1(x)| + |T_2(f)|)$$

$$\leq \frac{N}{\Gamma(p+q+1)} (t_2^{p+q} - t_1^{p+q}) + \frac{2|\mu|\xi}{\Gamma(q+1)} (t_2-t_1)^q + \frac{(t_2^{q+1} - t_1^{q+1})}{(1-\omega)} \left\{ \xi \frac{|\mu|}{\Gamma(q+1)} \right.$$

$$+ \frac{N}{\Gamma(p+q+1)} + N \frac{\sum_{i=1}^n |\alpha_i| \eta_i^{p+q}}{\Gamma(p+q+1)} + \xi \frac{|\mu| \sum_{i=1}^n |\alpha_i| \eta_i^q}{\Gamma(q+1)} + \xi \sum_{i=1}^n |\beta_i| |\eta_i| \right\}$$

Similarly,

$$|({}^cD^r Tx)(t_2) - {}^cD^r(Tx)(t_1)| \leq \frac{N}{\Gamma(p+q-r+1)} (t_2^{p+q-r} - t_1^{p+q-r}) + \frac{2|\mu|\xi}{\Gamma(q-r+1)} (t_2-t_1)^{q-r}$$

$$+ \frac{\Gamma(q+2)(t_2^{q-r+1} - t_1^{q-r+1})}{\Gamma(q-r+2)(1-\omega)} \left\{ \xi \frac{|\mu|}{\Gamma(q+1)} + \frac{N}{\Gamma(p+q+1)} + N \frac{\sum_{i=1}^n |\alpha_i| \eta_i^{p+q}}{\Gamma(p+q+1)} \right.$$

$$+ \xi \frac{|\mu| \sum_{i=1}^n |\alpha_i| \eta_i^q}{\Gamma(q+1)} + \xi \sum_{i=1}^n |\beta_i| |\eta_i| \right\}$$

Observe that $(t_2^{p+q} - t_1^{p+q})$, $(t_2^{q+1} - t_1^{q+1})$, $(t_2-t_1)^q$, $(t_2^{p+q-r} - t_1^{p+q-r})$, $(t_2^{q-r+1} - t_1^{q-r+1})$, and $(t_2-t_1)^{q-r}$ approach uniformly zero as t_1 approaches t_2. Then, the operator T is equicontinuous, and we get that the operator T is uniformly bounded since $T\mathcal{B}_\xi \subset \mathcal{B}_\xi$. Therefore, the Arzela–Ascoli theorem leads to that the operator being completely continuous. Hence, the Schauder fixed point theorem ensures the existence of the solution for problem Equations (1) and (2). □

The second result is discussing the existence of the solution by using Krasnoselskii–Zabreiko's fixed point theorem:

Lemma 4 ([41]). *Let $\mathcal{W}$ be a Banach space. Suppose that $\mathfrak{F} : \mathcal{W} \to \mathcal{W}$ is a completely continuous mapping and $\mathfrak{G} : \mathcal{W} \to \mathcal{W}$ is a bounded linear mapping such that 1 is not an eigenvalue of $\mathfrak{G}$ and:*

$$\lim_{\|x\| \to \infty} \frac{\|\mathfrak{F}x - \mathfrak{G}x\|}{\|x\|} = 0.$$

Then, $\mathfrak{F}$ has a fixed point in $\mathcal{W}$.

Theorem 2. *Assume that G_1, G_3, and G_4 hold. Then, the boundary value problem Equations (1) and (2) have a solution if $\|\kappa\|\Theta + Y < 1$ where Θ and Y are defined as in Equations (7) and (8), respectively.*

Proof. Let $f(t, x(t), {}^cD^r x(t)) = \kappa(t)(x(t) + {}^cD^r x(t))$, and consider the linear operator:

$$(Fx)(t) = \frac{1}{\Gamma(q+p)} \int_0^t (t-s)^{q+p-1} \kappa(s)(x(s) + {}^cD^r x(s))ds - \frac{\mu}{\Gamma(q)} \int_0^t (t-s)^{q-1} x(s) ds$$
$$+ \frac{t^{q+1}}{1-\omega} [T_1(x) + T_2(\kappa(x + {}^cD^r x))]$$

and its fractional derivative:

$${}^cD^r(Fx)(t) = \frac{1}{\Gamma(q+p-r-1)} \int_0^t (t-s)^{q+p-1} \kappa(s)(x(s) + {}^cD^r x(s))ds$$
$$- \frac{\mu}{\Gamma(q-r)} \int_0^t (t-s)^{q-r-1} x(s) ds$$
$$+ \frac{\Gamma(q+2)t^{q+1-r}}{(1-\omega)\Gamma(q-r+2)} [T_1(x) + T_2(\kappa(x + {}^cD^r x))].$$

Now, we show that one is not an eigenvalue of $Fx(t)$ and ${}^cD^r Fx(t)$. Use the proof by contradiction. Suppose that one is an eigenvalue of $Fx(t)$, and ${}^cD^r Fx(t)$ using Equations (11) and (12), we can deduce that:

$$\|Fx\| \leq \|x\|_X \left\{ \frac{\|\kappa\|}{\Gamma(p+q+1)} + \frac{|\mu|}{\Gamma(q+1)} + \frac{|\mu|}{\Gamma(q+1)(1-\omega)} + \frac{\|\kappa\|}{\Gamma(p+q+1)(1-\omega)} \right.$$
$$\left. + \frac{\|\kappa\| \sum_{i=1}^n |\alpha_i| \eta_i^{p+q}}{\Gamma(p+q+1)} + \frac{|\mu| \sum_{i=1}^n \eta_i^q}{\Gamma(q+1)(1-\omega)} + \sum_{i=1}^n |\beta_i||\eta_i| \right\}$$

and:

$$\|{}^cD^r Fx\| \leq \|x\|_X \left\{ \frac{\|\kappa\|}{\Gamma(p+q-r+1)} + \frac{|\mu|}{\Gamma(q-r+1)} + \frac{|\mu|\Gamma(q+2)}{\Gamma(q-r+1)\Gamma(q+1)(1-\omega)} \right.$$
$$+ \frac{\|\kappa\|\Gamma(q+2)}{\Gamma(q-r+1)\Gamma(p+q+1)(1-\omega)} + \frac{\|\kappa\|\Gamma(q+2)\sum_{i=1}^n |\alpha_i|\eta_i^{p+q}}{\Gamma(q-r+2)\Gamma(p+q+1)}$$
$$\left. + \frac{|\mu|\Gamma(q+2)|\sum_{i=1}^n \eta_i^q}{\Gamma(q-r+2)\Gamma(q+1)(1-\omega)} + \frac{\Gamma(q+2)\sum_{i=1}^n |\beta_i||\eta_i|}{\Gamma(q-r+2)(1-\omega)} \right\}.$$

This implies:

$$\|Fx\|_X = \max |(Fx)(t)| + \max |({}^cD^r(Fx)(t))|$$
$$\leq \|x\|_X \{\|\kappa\|\Theta + Y\} < \|x\|_X.$$

This is a contradiction, because our supposition that one is an eigenvalue of the operator $\|Fx\|_X$ is the wrong assumption. Hence, one is not an eigenvalue of $\|Fx\|_X$. The operator $\|Tx\|_X$ is uniformly bounded and equicontinuous as in Theorem 1.

Now, we prove $\frac{\|Tx-Fx\|_X}{\|x\|_X} \to 0$ as $\|x\|_X \to \infty$, then:

$$\|Tx - Fx\| \leq \frac{1}{\Gamma(q+p)} \int_0^t (t-s)^{q+p-1} |f(s, x(s) \, ^cD^r x(s)) - \kappa(s)(x(s) + \, ^c D^r x(s))| ds$$

$$+ \frac{1}{|1-\omega|} |T_2(f) - T_2(\kappa(x + \, ^c D^r x))|$$

$$\leq \frac{1}{\Gamma(q+p)} \int_0^t (t-s)^{q+p-1} \left| \frac{f(s, x(s) \, ^c D^r x(s))}{x(s) + \, ^c D^r x(s)} - \kappa(s) \right| (|x(s)| + |^c D^r x(s)|) ds$$

$$+ \frac{1}{|1-\omega|} \left[\frac{1}{\Gamma(p+q)} \int_0^1 (1-s)^{q+p-1} \left| \frac{f(s, x(s) \, ^c D^r x(s))}{x(s) + \, ^c D^r x(s)} - \kappa(s) \right| (|x(s)| + |^c D^r x(s)|) ds \right.$$

$$+ \sum_{i=1}^n \frac{|\alpha_i|}{\Gamma(p+q)} \int_0^{\eta_i} (\eta_i - s)^{q+p-1} \left| \frac{f(s, x(s) \, ^c D^r x(s))}{x(s) + \, ^c D^r x(s)} - \kappa(s) \right| (|x(s)| + |^c D^r x(s)|) ds \right]$$

$$\leq \left\{ \frac{1}{\Gamma(p+q+1)} + \frac{1}{\Gamma(p+q+1)(1-\omega)} + \frac{\sum_{i=1}^n |\alpha_i| \eta_i^{p+q}}{\Gamma(p+q+1)|1-\omega|} \right\} \left| \frac{f(t, x(t) \, ^c D^r x(t))}{x(t) + \, ^c D^r x(t)} - \kappa(t) \right| \|x\|_X$$

and similarly,

$$\|\, ^c D^r Tx - \, ^c D^r Fx\| \leq \left\{ \frac{1}{\Gamma(p+q-r+1)} + \frac{\Gamma(q+2)}{\Gamma(q-r+2)\Gamma(p+q+1)|1-\omega|} \right.$$

$$\left. + \frac{\Gamma(q+2) \sum_{i=1}^n |\alpha_i| \eta_i^{p+q}}{\Gamma(q-r+2)\Gamma(p+q+1)|1-\omega|} \right\} \left| \frac{f(t, x(t) D^r x(t))}{x(t) + \, ^c D^r x(t)} - \kappa(t) \right| \|x\|_X$$

Therefore,

$$\frac{\|Tx - Fx\|_X}{\|x\|_X} = \frac{\|(Tx)(t) - (Fx)(t)\|}{\|x\|_X} + \frac{\|\, ^c D^r(Tx)(t) - \, ^c D^r(Fx)(t)\|}{\|x\|_X}$$

$$\leq \left\{ \frac{1}{\Gamma(p+q+1)} + \frac{1 + \sum_{i=1}^n |\alpha_i| \eta_i^{p+q}}{(1-\omega)\Gamma(p+q+1)} + \frac{1}{\Gamma(p+q-r+1)} + \frac{\Gamma(q+2)(1 + \sum_{i=1}^n |\alpha_i| \eta_i^{p+q})}{(1-\omega)\Gamma(q-r+2)\Gamma(p+q+1)} \right\}$$

$$\times \left| \frac{f(t, x(t) \, ^c D^r x(t))}{x(t) + \, ^c D^r x(t)} - \kappa(t) \right|$$

By assumption G4,

$$\lim_{\|x\| \to \infty} \frac{\|Tx - Fx\|_X}{\|x\|_X} = 0$$

Then, Krasnoselskii–Zabreiko's fixed point theorem leads to the existence of a solution for the boundary value problem Equations (1) and (2). □

We present now the Banach contraction principle to prove the uniqueness of the solution for problem Equations (1) and (2).

Theorem 3. *Assume that G_1 and G_5 hold. The boundary value problem Equations (1) and (2) have a unique solution if $\sigma < 1$ where:*

$$\sigma = L\Theta + Y$$

and Θ and Y are defined as in Equations (7) and (8), respectively.

Proof. Using the assumption G_5, we have:

$$|(Tx)(t) - (T\hat{x})(t)| \leq \frac{1}{\Gamma(q+p)} \int_0^t (t-s)^{q+p-1} |f(s, x(s), {}^cD^r x(s)) - f(s, \hat{x}(s), {}^cD^r \hat{x}(s))| ds$$

$$+ \frac{|\mu|}{\Gamma(q)} \int_0^t (t-s)^{q-1} |x(s) - \hat{x}(s)| ds$$

$$+ \frac{1}{|1-\omega|} \left(|T_1(x) - T_1(\hat{x})| + |T_2(f(s, x(s), {}^cD^r x(s))) - T_2(f(s, \hat{x}(s), {}^cD^r \hat{x}(s)))| \right)$$

$$\leq L\|x - \hat{x}\|_X \left\{ \frac{t^{p+q}}{\Gamma(p+q+1)} + \frac{1}{|1-\omega|\Gamma(p+q+1)} + \frac{\sum_{i=1}^n |\alpha_i|\eta_i^{p+q}}{|1-\omega|\Gamma(p+q+1)} \right\}$$

$$+ \|x - \hat{x}\|_X \left\{ \frac{|\mu|t^q}{\Gamma(q+1)} + \frac{t^{q+1}|\mu|}{|1-\omega|\Gamma(q+1)} + \frac{t^{q+1}|\mu|}{|1-\omega|\Gamma(q+1)} \sum_{i=1}^n |\alpha_i|\eta_i^q + \frac{1}{|1-\omega|} \sum_{i=1}^n |\beta_i||\eta_i| \right\}$$

$$\leq (L\Theta_{p,0} + Y_0)\|x - \hat{x}\|_X$$

likewise,

$$|{}^cD^r(Tx)(t) - {}^cD^r(T\hat{x})(t)| \leq (L\Theta_{p,r} + Y_r)\|x - \hat{x}\|_X$$

Hence,

$$\|Tx - T\hat{x}\|_X = \max_{t \in [0,1]} |(Tx)(t) - (T\hat{x})(t)| + \max_{t \in [0,1]} |({}^cD^r(Tx)(t)) - ({}^cD^r(T\hat{x})(t))|$$

$$\leq (L\Theta_{p,0} + Y_0)\|x - \hat{x}\|_X + (L\Theta_{p,r} + Y_r)\|x - \hat{x}\|_X$$

$$\leq (L\Theta + Y)\|x - \hat{x}\|_X \leq \sigma \|x - \hat{x}\|_X$$

Since $\sigma < 1$, then the operator Tx is a contraction. Therefore, from the contraction mapping principle, the boundary value problem Equations (1) and (2) have a unique solution on $[0,1]$. □

4. Example

Example 1. *Consider the following boundary value problem:*

$$\begin{cases} {}^cD^{\frac{9}{8}} \left({}^cD^{\frac{5}{8}} + \frac{1}{10} \right) x(t) = f(t, x(t), D^{\frac{3}{8}} x(t)), & 0 < t < 1 \\ x(0) = 0, \quad {}^cD^{\frac{5}{8}} x(0) = 0, \\ x(1) = \frac{1}{4} x(\frac{1}{3}) + \frac{1}{2} x(\frac{1}{9}) + \frac{1}{5} \int_0^{1/3} x(s) ds + \frac{2}{5} \int_0^{1/9} x(s) ds. \end{cases} \quad (13)$$

We choose $p = 9/8$, $q = 5/8$, $r = 3/8$, $\alpha_i = \frac{i}{4}$, $\beta_i = \frac{i}{7}$, $\eta_i = \frac{1}{3^i}$, $(i = 1, 2)$, and $\mu = 1/4$. Define the continuous function by:

$$f(t, x, y) = \frac{e^t \sin(\pi t)}{(2+t)^3} + \frac{t \cos^2 \pi t}{(3-t)^4} (x + y)$$

Observe that the function f is continuous and $f(t, 0, 0) = \frac{e^t \sin(\pi t)}{(2+t)^3} \neq 0$ on $(0,1)$, which means that the assumptions G_1 and G_3 hold. Now, we have:

$$\frac{f(t, x, y)}{x+y} = \frac{e^t \sin(\pi t)}{(2+t)^3(x+y)} + \frac{t \cos^2 \pi t}{(3-t)^4}$$

which implies that:

$$\lim_{\|x\|_X \to \infty} \frac{f(t, x, y)}{x+y} = \frac{t \cos^2 \pi t}{(3-t)^4} = \kappa(t) \quad \text{and} \quad \|\kappa\| = \frac{1}{16}.$$

Thus, we can calculate $\|\kappa\|\Theta + Y \sim \frac{1}{16}(3.01664) + 0.75232 = 0.94086 < 1$. Therefore, by Theorem 2, the boundary value problem Equations (1) and (2) have a solution in $[0, 1]$.

Furthermore, it is clear that the function f satisfies the assumption G_5 with $L = 1/16$ and $\sigma = L\Theta + Y \sim 0.94086 < 1$. Then, by Theorem 3, the boundary value problem Equations (1) and (2) have a unique solution in $[0, 1]$.

5. Conclusions

The existence and uniqueness of the solution for the fractional nonlinear Langevin equation of two different fractional orders under the boundary conditions containing multi-point and multi-strip were studied. We found an equivalence of the problem by using the tools of fractional calculus and fixed point theorems. To examine our problem, we employed Krasnoselskii–Zabreiko, Schauder, and Banach contraction fixed point theorems. Our method was simple and appropriate for a diversity of real-world problems by choosing different forms of the function f in the Langevin equation. For instance, if $f = -\gamma(x(t))^c D^r x(t)) + \eta(x(t))\xi(t) + F(t, x(t))$ where $\gamma(x(t))$ is the position-dependent phenomenological fluid friction coefficient, $F(z, t)$ is the external force field, $\eta(x(t))$ is the intensity of the stochastic force, and $\xi(t)$ is a zero-mean Gaussian white noise term, then the model describes the fractional Markovian set of stochastic differential equations [42].

Author Contributions: Conceptualization, methodology, formal analysis, A.S.; investigation, writing—original draft preparation, B.A. All authors have read and agreed to the published version of the manuscript.

Funding: This project was funded by the Deanship of Scientific Research (DSR), King Abdulaziz University, Jeddah. The authors, therefore, acknowledge with thanks DSR technical and financial support.

Conflicts of Interest: The authors declare no conflict of interest.

References

1. Coffey, W.T.; Kalmykov, Y.P.; Waldron, J.T. *The Langevin Equation: With Applications to Stochastic Problems in Physics, Chemistry and Electrical Engineering*; World Scientific: Singapore, 2004.
2. Lim, S.C.; Teo, L.P. The fractional oscillator process with two indices. *J. Phys. A Math. Theor.* **2009**, *42*, 065208. [CrossRef]
3. Lim, S.C.; Li, M.; Teo, L.P. Langevin equation with two fractional orders. *Phys. Lett.* **2008**, *372*, 6309–6320. [CrossRef]
4. Eab, C.H.; Lim, S.C. Fractional Langevin equation of distributed order. *arXiv* **2010**, arXiv:1010.3327.
5. Sandev, T.; Tomovski, Z. *Fractional Equations and Models: Theory and Applications*; Springer Nature: Geneva, Switzerland, 2019.
6. West, B.J.; Bologna, M.; Grigolini, P. *Physics of Fractal Operators*; Springer: New York, NY, USA, 2003.
7. Kobelev, V.; Romanov, E. Fractional Langevin equation to describe anamalous diffusion Prog. Theor. Phys. Suppl. **2000**, *139*, 470-476. [CrossRef]
8. Sandev, T.; Metzler, R.; Tomovski, Z. Correlation functions for the fractional generalized Langevin equation in the presence of internal and external noise. *J. Math. Phys.* **2014**, *55*, 023301. [CrossRef]
9. Sandev, T.; Metzler, R.; Tomovski, Z. Velocity and displacement correlation functions for fractional generalized Langevin equations. *Fract. Calc. Appl. Anal.* **2012**, *15*, 426. [CrossRef]
10. West, B.J. Fractal physiology and the fractional calculus: A perspective. *Front. Physiol.* **2010**, *1*, 12. [CrossRef]
11. Kolmogorov, A.N. Wienersche Spiralen und einige andere interessante Kurvenim Hilbertschen Raum. *Dokl. Acad. Sci. USSR* **1940**, *26*, 115.
12. Mandelbrot, B.B.; van Ness, J.W. Fractional Brownian motions, fractional noises and applications. *SIAM Rev.* **1968**, *10*, 422. [CrossRef]
13. Mainardi, F.; Pironi, P. The fractional Langevin equation:Brownian motion revisted. *Extracta Math.* **1996**, *10*, 140–154.
14. Camargo, R.F.; Chiacchio, A.O.; Charnet, R.; Oliveira, E.C. Solution of the fractional Langevin equation and the Mittag-Leffler functions. *J. Math. Phys.* **2009**, *50*, 063507. [CrossRef]

15. Vinales, A.D.; Desposito, M.A. Anomalous diffusion induced by a Mittag-Leffler correlated noise. *Phys. Rev. E* **2007**, *75*, 042102. [CrossRef] [PubMed]
16. Guo, P.; Zeng, C.; Li, C.; Chen, Y. Numerics for the fractional Langevin equation driven by the fractional Brownian motion. *Fract. Calc. Appl. Anal.* **2013**, *16*, 123–141. [CrossRef]
17. Guo, P.; Li, C.P.; Zeng, F.H. Numerical simulation of the fractional Langevin equation. *Therm. Sci.* **2012**, *16*, 357–363. [CrossRef]
18. Zhao, J.; Lu, P.; Liu, Y. Existence and Numerical Simulation of Solutions forFractional Equations Involving Two Fractional Orders withNonlocal Boundary Conditions. *J. Appl. Math.* **2013**, *2013*, 268347. [CrossRef]
19. Mahmudov, N.I. Fractional Langevin type delay equations with two fractional derivatives. *Appl. Math. Lett.* **2020**, *103*, 106215. [CrossRef]
20. Baghani, H.; Nieto, J.J. On fractional Langevin equation involving two fractional orders in different intervals. *Nonlinear Anal. Model. Control.* **2019**, *24*, 884–897. [CrossRef]
21. Zhai, C.; Li, P. Nonnegative Solutions of Initial Value Problems for Langevin Equations Involving Two Fractional Orders. *Mediterr. J. Math.* **2018**, *15*, 164. [CrossRef]
22. Baghani, H. Existence and uniqueness of solutions to fractional Langevin equations involving two fractional orders. *J. Fixed Point Theory Appl.* **2018**, *20*, 63. [CrossRef]
23. Fazli, H.; Nieto, J.J. Fractional Langevin equation with anti-periodic boundary conditions. *Chaos Solitons Fractals* **2018**, *114*, 332–337. [CrossRef]
24. Zhai, C.; Li, P.; Li, H. Single upper-solution or lower-solution method for Langevin equations with two fractional orders. *Adv. Differ. Equ.* **2018**, *360*, 1–10. [CrossRef]
25. Baghani, O. On fractional Langevin equation involving two fractional orders. *Commun. Nonlinear Sci. Numer. Simulat.* **2017**, *42*, 675–681. [CrossRef]
26. Cetin, E.; Topa, F.S. Existence Results for Solutions of Integral Boundary Value Problems on Time Scales. *Abstr. Appl. Anal.* **2013**, *708734*, 7. [CrossRef]
27. Salem, A.; Alzahrani, F.; Alnegga, M. Coupled System of Non-linear Fractional Langevin Equations with Multi-point and Nonlocal Integral Boundary Conditions. *Math. Probl. Eng.* **2020**, *7345658*, 15.
28. Salem, A.; Alzahrani, F.; Almaghamsi, L. Fractional Langevin equation with nonlocal integral boundary condition. *Mathematics* **2019**, *7*, 402. [CrossRef]
29. Zhou, Z.; Qiao, Y. Solutions for a class of fractional Langevin equations with integral and anti-periodic boundary conditions. *Bound. Value Probl.* **2018**, *2018*, 152. [CrossRef]
30. Salem, A.; Alzahrani, F.; Alghamdi, B. Langevin equation involving two fractional orders with three-point boundary conditions. *Differ. Integral Equ.* **2020**, *33*, 163–180.
31. Salem, A.; Alghamdi, B. Multi-Point and Anti-Periodic Conditions for Generalized Langevin Equation with Two Fractional Orders. *Fractal Fract.* **2019**, *3*, 51. [CrossRef]
32. Derbazi1, C.; Hammouche, H.; Benchohra, M.; Zhou, Y. Fractional hybrid differential equations withthree-point boundary hybrid conditions. *Adv. Differ. Equ.* **2019**, *2019*, 125. [CrossRef]
33. Lv, Z.-W. Existence of Positive Solution for Fractional Differential Systems with Multi-point Boundary Value Conditions. *J. Funct. Spaces Vol.* **2020**, *9520430*, 9.
34. Sandin, T.R. The jerk. *Phys. Teach.* **1998**, *28*, 36–40. [CrossRef]
35. Schot, S.H. Jerk: The time rate of change of acceleration. *Am. J. Phys.* **1978**, *46*, 1090–1094. [CrossRef]
36. Schot, S.H. Aberrancy: Geometry of the Third Derivative. *Math. Mag.* **1978**, *51*, 259–275. [CrossRef]
37. Granas, A.; Dugundji, J. *Fixed Point Theory*; Springer: New York, NY, USA, 2003.
38. Kilbas, A.A.; Srivastava, H.M.; Trujillo, J.J. *Theory and Applications of Fractional Differential Equations*; Elsevier: Amsterdam, The Netherlands, 2006.
39. Podlubny, I. Fractional Differential Equations. In *Mathematics in Science and Engineering*; Academic Press: New York, NY, USA, 1999; Volume 198.
40. Su, X. Boundary value problem for a coupled system of nonlinear fractional differential equations. *Appl. Math. Lett.* **2009**, *22*, 64–69. [CrossRef]

41. Krasnoselski, M.A.; Zabreiko, P.P. *Geometrical Methods of Nonlinear Analysis*; Springer: New York, NY, USA, 1984.
42. Olivares-Rivas, W.; Colmenares, P.J. The generalized Langevin equation revisited: Analytical expressions for the persistence dynamics of a viscous fluid under a time dependent external force. *Phys. A* **2016**, *458*, 76–94. [CrossRef]

© 2020 by the authors. Licensee MDPI, Basel, Switzerland. This article is an open access article distributed under the terms and conditions of the Creative Commons Attribution (CC BY) license (http://creativecommons.org/licenses/by/4.0/).

Article

A Stochastic Fractional Calculus with Applications to Variational Principles

Houssine Zine [†,‡] and Delfim F. M. Torres [*,‡]

Center for Research and Development in Mathematics and Applications (CIDMA), Department of Mathematics, University of Aveiro, 3810-193 Aveiro, Portugal; zinehoussine@ua.pt
* Correspondence: delfim@ua.pt
† This research is part of first author's Ph.D. project, which is carried out at the University of Aveiro under the Doctoral Program in Applied Mathematics of Universities of Minho, Aveiro, and Porto (MAP).
‡ These authors contributed equally to this work.

Received: 19 May 2020; Accepted: 30 July 2020; Published: 1 August 2020

Abstract: We introduce a stochastic fractional calculus. As an application, we present a stochastic fractional calculus of variations, which generalizes the fractional calculus of variations to stochastic processes. A stochastic fractional Euler–Lagrange equation is obtained, extending those available in the literature for the classical, fractional, and stochastic calculus of variations. To illustrate our main theoretical result, we discuss two examples: one derived from quantum mechanics, the second validated by an adequate numerical simulation.

Keywords: fractional derivatives and integrals; stochastic processes; calculus of variations

MSC: 26A33; 49K05; 60H10

---

1. Introduction

A stochastic calculus of variations, which generalizes the ordinary calculus of variations to stochastic processes, was introduced in 1981 by Yasue, generalizing the Euler–Lagrange equation and giving interesting applications to quantum mechanics [1]. Recently, stochastic variational differential equations have been analyzed for modeling infectious diseases [2,3], and stochastic processes have shown to be increasingly important in optimization [4].

In 1996, fifteen years after Yasue's pioneer work [1], the theory of the calculus of variations evolved in order to include fractional operators and better describe non-conservative systems in mechanics [5]. The subject is currently under strong development [6]. We refer the interested reader to the introductory book [5] and to [7–9] for numerical aspects on solving fractional Euler–Lagrange equations. For applications of fractional-order models and variational principles in epidemics, biology, and medicine, see [10–14] and references therein.

Given the importance of both stochastic and fractional calculi of variations, it seems natural to join the two subjects. That is the main goal of our current work, i.e., to introduce a stochastic-fractional calculus of variations. For that, we start our work by introducing new definitions: left and right stochastic fractional derivatives and integrals of Riemann–Liouville and Caputo types for stochastic processes of second order, as a deterministic function resulting from the intuitive action of the expectation, on which we can compute its fractional derivative several times to obtain additional results that generalize analogous classical relations. Our definitions differ from those already available in the literature by the fact that they are applied on second order stochastic processes, whereas known definitions, for example, those in [15–18], are defined only for mean square continuous second order stochastic process, which is a short family of operators. Moreover, available results in the literature have not used the expectation, which we claim to be more natural, easier to handle and estimate, when applied to fractional derivatives by different methods of approximation, like those developed and cited in [7]. More than different, our definitions are well posed and lead to numerous results generalizing those in the literature, like integration by parts and Euler–Lagrange variational equations.

The paper is organized as follows. In Section 2, we introduce the new stochastic fractional operators. Their fundamental properties are then given in Section 3. In particular, we prove stochastic fractional formulas of integration by parts (see Lemma 1). Then, in Section 4, we consider the basic problem of the stochastic fractional calculus of variations and obtain the stochastic Riemann–Liouville and Caputo fractional Euler–Lagrange equations (Theorems 1 and 2, respectively). Section 5 gives two illustrative examples. We end with Section 6 on conclusions and future perspectives.

2. The Stochastic Fractional Operators

Let (Ω, F, P) be a probabilistic space, where Ω is a nonempty set, F is a σ-algebra of subsets of Ω, and P is a probability measure defined on Ω. A mapping X from an open time interval I into the Hilbert space $H = L_2(\Omega, P)$ is a stochastic process of second order in $\mathbb{R}$. We introduce the stochastic fractional operators by composing the classical fractional operators with the expectation E.

In what follows, the classical fractional operators are denoted using standard notations [19]: ${}_aD_t^\alpha$ and ${}_tD_b^\alpha$ denote the left and right Riemann–Liouville fractional derivatives of order α; ${}_aI_t^\alpha$ and ${}_tI_b^\alpha$ the left and right Riemann–Liouville fractional integrals of order α; while the left and right Caputo fractional derivatives of order α are denoted by ${}_a^CD_t^\alpha$ and ${}_t^CD_b^\alpha$, respectively. The new stochastic operators add to the standard notations an 's' for "stochastic".

Definition 1 (Stochastic fractional operators). *Let X be a stochastic process on $[a,b] \subset I$, $\alpha > 0$, $n = [\alpha] + 1$, such that $E(X(t)) \in AC^n([a,b] \to \mathbb{R})$ with AC the class of absolutely continuous functions. Then,*

(D1) *the left stochastic Riemann–Liouville fractional derivative of order α is given by*

$$\begin{aligned}{}_a^sD_t^\alpha X(t) &= {}_aD_t^\alpha[E(X_t)] \\ &= \frac{1}{\Gamma(n-\alpha)}\left(\frac{d}{dt}\right)^n \int_a^t (t-\tau)^{n-1-\alpha} E(X_\tau) d\tau, \quad t > a;\end{aligned}$$

(D2) *the right stochastic Riemann–Liouville fractional derivative of order α by*

$$\begin{aligned}{}_t^sD_b^\alpha X(t) &= {}_tD_b^\alpha[E(X_t)] \\ &= \frac{1}{\Gamma(n-\alpha)}\left(\frac{-d}{dt}\right)^n \int_t^b (\tau-t)^{n-1-\alpha} E(X_\tau) d\tau, \quad t < b;\end{aligned}$$

(D3) *the left stochastic Riemann–Liouville fractional integral of order α by*

$$\begin{aligned}{}_a^sI_t^\alpha X(t) &= {}_aI_t^\alpha[E(X_t)] \\ &= \frac{1}{\Gamma(\alpha)} \int_a^t (t-\tau)^{\alpha-1} E(X_\tau) d\tau, \quad t > a;\end{aligned}$$

(D4) *the right stochastic Riemann–Liouville fractional integral of order α by*

$$\begin{aligned}{}_t^sI_b^\alpha X(t) &= {}_tI_b^\alpha[E(X_t)] \\ &= \frac{1}{\Gamma(\alpha)} \int_t^b (\tau-t)^{\alpha-1} E(X_\tau) d\tau, \quad t < b;\end{aligned}$$

(D5) *the left stochastic Caputo fractional derivative of order α by*

$$\begin{aligned}{}_a^{sC}D_t^\alpha X(t) &= {}_a^CD_t^\alpha[E(X_t)] \\ &= \frac{1}{\Gamma(n-\alpha)} \int_a^t (t-\tau)^{n-1-\alpha} E(X(\tau))^{(n)} d\tau; \quad t > a.\end{aligned}$$

(D6) and the right stochastic Caputo fractional derivative of order α by

$$\begin{aligned}{}_t^{sC}D_b^\alpha X(t) &= {}_t^{C}D_b^\alpha [E(X_t)] \\ &= \frac{(-1)^n}{\Gamma(n-\alpha)} \int_t^b (\tau-t)^{n-1-\alpha} E(X(\tau))^{(n)} d\tau, \quad t < b.\end{aligned}$$

Remark 1. *The stochastic processes $X(t)$ used along the manuscript can be of any type satisfying the announced conditions of existence of the novel stochastic fractional operators. For example, we can consider Levy processes as a particular case, provided one considers some intervals where $E(X(t))$ is sufficiently smooth [20].*

As we shall prove in the following sections, the new stochastic fractional operators just introduced provide a rich calculus with interesting applications.

3. Fundamental Properties

Several properties of the classical fractional operators, like boundedness or linearity, also hold true for their stochastic counterparts.

Proposition 1. *If $t \to E(X_t) \in L_1([a,b])$, then ${}_a^s I_t^\alpha (X_t)$ is bounded.*

Proof. The property follows easily from definition (D3):

$$|{}_a^s I_t^\alpha (X_t)| = \left| \frac{1}{\Gamma(\alpha)} \int_a^t (t-\tau)^{\alpha-1} E(X_\tau) d\tau \right| \le k \|E(X_t)\|_1,$$

which shows the intended conclusion. □

Proposition 2. *The left and right stochastic Riemann–Liouville and Caputo fractional operators given in Definition 1 are linear operators.*

Proof. Let c and d be real numbers and assume that ${}_a^s D_t^\alpha X_t$ and ${}_a^s D_t^\alpha Y_t$ exist. It is easy to see that ${}_a^s D_t^\alpha (c \cdot X_t + d \cdot Y_t)$ also exists. From Definition 1 and by linearity of the expectation and the linearity of the classical/deterministic fractional derivative operator, we have

$$\begin{aligned}{}_a^s D_t^\alpha (c \cdot X_t + d \cdot Y_t) &= {}_a D_t^\alpha E(c \cdot X_t + d \cdot Y_t) \\ &= c \cdot {}_a D_t^\alpha E(X_t) + d \cdot {}_a D_t^\alpha E(Y_t) \\ &= c \cdot {}_a^s D_t^\alpha (X_t) + d \cdot {}_a^s D_t^\alpha (Y_t).\end{aligned}$$

The linearity of the other stochastic fractional operators is obtained in a similar manner. □

Our next proposition involves both stochastic and deterministic operators. Let $\mathcal{O} \in \{D, I, {}^C D\}$. Recall that if ${}_a^s \mathcal{O}_t^\beta$ is a left stochastic fractional operator of order β, then ${}_a \mathcal{O}_t^\beta$ is the corresponding left classical/deterministic fractional operator of order β; similarly for right operators.

Note that the proofs of Propositions 3 and 4 and Lemma 1 are not hard to prove in the sense that they are based on well-known results available for deterministic fractional derivatives (observe that $E(X(t))$ is deterministic).

Proposition 3. Assume that ${}^s_aI^\beta_t X_t$, ${}^s_tI^\beta_b X_t$, ${}^s_aI^\alpha_t X_t$, ${}_aD^\alpha_t[{}^s_aI^\alpha_t X_t]$, ${}_aI^\alpha_t\left[{}^s_aI^\beta_t X_t\right]$ and ${}_tI^\alpha_b\left[{}^s_tI^\beta_b X_t\right]$ exist. The following relations hold:

$${}_aI^\alpha_t\left[{}^s_aI^\beta_t X_t\right] = {}^s_aI^{\alpha+\beta}_t X_t,$$

$${}_tI^\alpha_b\left[{}^s_tI^\beta_b X_t\right] = {}^s_tI^{\alpha+\beta}_b X_t,$$

$${}_aD^\alpha_t[{}^s_aI^\alpha_t X_t] = E(X_t).$$

Proof. Using Definition 1 and well-known properties of the deterministic Riemann–Liouville fractional operators [21], one has

$$\begin{aligned}
{}_aI^\alpha_t\left[{}^s_aI^\beta_t X_t\right] &= {}_aI^\alpha_t\left[{}_aI^\beta_t E(X_t)\right] \\
&= {}_aI^{\alpha+\beta}_t E(X_t) \\
&= {}^s_aI^{\alpha+\beta}_t X_t.
\end{aligned}$$

The second and third equalities are easily proved in a similar manner. □

Proposition 4. Let $\alpha > 0$. If $E(X_t) \in L_\infty(a,b)$, then

$${}^C_aD^\alpha_t[{}^s_aI^\alpha_t X_t] = E(X_t)$$

and

$${}^C_tD^\alpha_b[{}^s_tI^\alpha_b X_t] = E(X_t).$$

Proof. Using Definition 1 and well-known properties of the deterministic Caputo fractional operators [21], we have

$$\begin{aligned}
{}^C_aD^\alpha_t[{}^s_aI^\alpha_t X_t] &= {}^C_aD^\alpha_t[{}_aI^\alpha_t E(X_t)] \\
&= E(X_t).
\end{aligned}$$

The second formula is shown with the same argument. □

Formulas of integration by parts play a fundamental role in the calculus of variations and optimal control [22,23]. Here we make use of Lemma 1 to prove in Section 4 a stochastic fractional Euler–Lagrange necessary optimality condition.

Lemma 1 (Stochastic fractional formulas of integration by parts). *Let $\alpha > 0$, $p,q \geq 1$, and $\frac{1}{p} + \frac{1}{q} \leq 1 + \alpha$ ($p \neq 1$ and $q \neq 1$ in the case where $\frac{1}{p} + \frac{1}{q} = 1 + \alpha$).*

(i) If $E(X_t) \in L_p(a,b)$ and $E(Y_t) \in L_q(a,b)$ for every $t \in [a,b]$, then

$$E\left(\int_a^b (X_t){}^s_aI^\alpha_t Y_t dt\right) = E\left(\int_a^b (Y_t){}^s_tI^\alpha_b X_t dt\right).$$

(ii) If $E(Y_t) \in {}_tI^\alpha_b(L_p)$ and $E(X_t) \in {}_aI^\alpha_t(L_q)$ for every $t \in [a,b]$, then

$$E\left(\int_a^b (X_t)({}^s_aD^\alpha_t Y_t) dt\right) = E\left(\int_a^b (Y_t)({}^s_tD^\alpha_b X_t) dt\right).$$

(iii) For the stochastic Caputo fractional derivatives, one has

$$E\left[\int_a^b (X_t)({}^{sC}_aD^\alpha_t Y_t) dt\right] = E\left[\int_a^b (Y_t)({}^s_tD^\alpha_b X_t) dt\right] + E\left[({}^s_tI^{1-\alpha}_b X_t)\cdot Y_t\right]_a^b$$

and
$$E\left[\int_a^b (X_t)(^{sC}_t D^\alpha_b Y_t)dt\right] = E\left[\int_a^b (Y_t)(^s_a D^\alpha_t X_t)dt\right] - E\left[(^s_a I^{1-\alpha}_t X_t)\cdot Y_t\right]_a^b$$

for $\alpha \in (0,1)$.

Proof. (i) We have

$$E\left(\int_a^b (X_t)^s_a I^\alpha_t Y_t dt\right) = \int_a^b E\left((X_t)^s_a I^\alpha_t Y_t\right) dt \quad \text{(by Fubini–Tonelli's theorem)}$$

$$= \int_a^b E\left((X_t)_a I^\alpha_t E(Y_t)\right) dt \quad \text{(by } (D_3)\text{)}$$

$$= \int_a^b E((X_t))_a I^\alpha_t E(Y_t) dt \quad \text{(the expectation is deterministic)}$$

$$= \int_a^b {}_t I^\alpha_b E(X_t) \cdot E(Y_t) dt \quad \text{(by fractional integration by parts)}$$

$$= E\left(\int_a^b {}^s_t I^\alpha_b (X_t)(Y_t) dt\right) \quad \text{(by Fubini–Tonelli's theorem)}.$$

(ii) With similar arguments as in item (i), we have

$$E\left(\int_a^b (X_t)^s_a D^\alpha_t Y_t dt\right) = \int_a^b E\left((X_t)^s_a D^\alpha_t Y_t\right) dt$$

$$= \int_a^b E\left((X_t)_a D^\alpha_t E(Y_t)\right) dt \quad \text{(by } (D_1)\text{)}$$

$$= \int_a^b E((X_t))_a D^\alpha_t E(Y_t) dt$$

$$= \int_a^b {}_t D^\alpha_b E(X_t) \cdot E(Y_t) dt$$

$$= E\left(\int_a^b {}^s_t D^\alpha_b (X_t)(Y_t) dt\right).$$

(iii) By using Caputo's fractional integration by parts formula we obtain that

$$E\left[\int_a^b (X_t)(^{sC}_a D^\alpha_t Y_t)dt\right] = \int_a^b E\left[(X_t)\right]\left(^C_a D^\alpha_t E\left[(Y_t)\right]\right) dt$$

$$= \int_a^b ({}_t D^\alpha_b E\left[(X_t)\right]\cdot E\left[(Y_t)\right]) dt + \left[({}_t I^{1-\alpha}_b E(X_t)\cdot E(Y_t)\right]_a^b$$

$$= \int_a^b ({}^s_t D^\alpha_b (X_t)\cdot E\left[(Y_t)\right]) dt + \left[({}_t I^{1-\alpha}_b E(X_t)\cdot E(Y_t)\right]_a^b$$

$$= E\left[\int_a^b ({}^s_t D^\alpha_b (X_t)\cdot (Y_t))dt\right] + E\left[({}_t I^{1-\alpha}_b E(X_t)\cdot (Y_t)\right]_a^b.$$

The first equality of (iii) is proved. By using a similar argument and applying the integration by parts formula associated with the right Caputo fractional derivative [21], we easily get the second equality of (iii). □

4. Stochastic Fractional Euler–Lagrange Equations

Let us denote by $C^1(I \to H)$ the set of second order stochastic processes X such that the left and right stochastic Riemann–Liouville fractional derivatives of X exist, endowed with the norm

$$\|X\| = \sup_{t \in I}\left(\|X(t)\|_H + |{}^s_a D^\alpha_t X(t)| + |{}^s_t D^\alpha_b X(t)|\right),$$

where $\|\cdot\|_H$ is the norm of H. Let $L \in C^1(I \times H \times \mathbb{R} \times \mathbb{R} \to \mathbb{R})$ and consider the following minimization problem:

$$J[X] = E\left(\int_a^b L\left(t, X(t), {}_a^s D_t^\alpha X(t), {}_t^s D_b^\alpha X(t)\right) dt\right) \longrightarrow \min \qquad (1)$$

subject to the boundary conditions

$$E(X(a)) = X_a, \quad E(X(b)) = X_b, \qquad (2)$$

where X verifies the above conditions and L is a smooth function. Taking into account the method used in [7] for the fractional setting, and according to stochastic fractional integration by parts given by our Lemma 1, we obtain the following necessary optimality condition for the fundamental problem (1)–(2) of the stochastic fractional calculus of variations.

Theorem 1 (The stochastic Riemann–Liouville fractional Euler–Lagrange equation). *If $J \in C^1(H \times \mathbb{R} \times \mathbb{R} \to \mathbb{R})$ and $X \in C^1(I \to H)$ is an F-adapted stochastic process on $[a, b]$ with $E(X(t)) \in AC([a, b])$ that is a minimizer of (1) subject to the fixed end points (2), then X satisfies the following stochastic fractional Euler–Lagrange equation:*

$$\frac{\partial L}{\partial X} + {}_t^s D_b^\alpha \left[\frac{\partial L}{\partial {}_a^s D_t^\alpha}\right] + {}_a^s D_t^\alpha \left[\frac{\partial L}{\partial {}_t^s D_b^\alpha}\right] = 0.$$

Proof. We have

$$J[X] = E\left(\int_a^b L(t, X(t), {}_a^s D_t^\alpha X(t), {}_t^s D_b^\alpha X(t) dt\right).$$

Assume that X^* is the optimal solution of problem (1)–(2). Set

$$X = X^* + \varepsilon \eta,$$

where η is an F-adapted stochastic process on $[a, b]$ in $C^1(I \to H)$. By linearity of the stochastic fractional derivatives (Proposition 2), we get

$${}_a^s D_t^\alpha X = {}_a^s D_t^\alpha X^* + \varepsilon \left({}_a^s D_t^\alpha \eta\right)$$

and

$${}_t^s D_b^\alpha X = {}_t^s D_b^\alpha X^* + \varepsilon \left({}_t^s D_b^\alpha \eta\right).$$

Consider now the following function:

$$j(\varepsilon) = E\left(\int_a^b L\left(t, X^* + \varepsilon \eta, {}_a^s D_t^\alpha X^* + \varepsilon \left({}_a^s D_t^\alpha \eta\right), {}_t^s D_b^\alpha X^* + \varepsilon \left({}_t^s D_b^\alpha \eta\right)\right) dt\right).$$

We deduce, by the chain rule, that

$$\frac{d}{dt} j(\varepsilon) \big|_{\varepsilon=0} = E\left(\int_a^b (\partial_2 L \cdot \eta + \partial_3 L \cdot {}_a^s D_t^\alpha \eta + \partial_4 L \cdot {}_t^s D_b^\alpha \eta) dt\right) = 0,$$

where $\partial_i L$ denotes the partial derivative of the Lagrangian L with respect to its ith argument. Using Lemma 1 of stochastic fractional integration by parts, we obtain

$$E\left(\int_a^b (\partial_2 L \cdot \eta + {}_t^s D_b^\alpha (\partial_3 L) \cdot \eta + {}_a^s D_t^\alpha (\partial_4 L) \cdot \eta) dt\right) = 0.$$

We claim that if Y is a stochastic process with continuous paths of second order such that

$$E\left[\int_a^b Y(t) \cdot \eta(t) dt\right] = 0$$

for any stochastic process with continuous paths η, then

$$Y = 0 \quad \text{almost surely (a.s., for short).}$$

Indeed, suppose that $Y(s) > 0$ a.s. for a certain $s \in (a,b)$. By continuity, $Y(t) > c > 0$ a.s. in a neighborhood of s, $a < s - r < s < s + r < b$, $r > 0$. Consider the process η such that $\eta(t) = 0$ a.s. on $[a, s-r] \cup [s+r, b]$ and $\eta(t) > 0$ a.s. on $(s-r, s+r)$, and $\eta(t) = 1$ a.s. on $\left(s - \frac{r}{2}, s + \frac{r}{2}\right)$. Then, $\int_a^b Y(t) \cdot \eta(t) dt \geq rc > 0$ a.s. Consequently, $E\left[\int_a^b Y(t) \cdot \eta(t) dt\right] > 0$, which completes the proof of our claim. Taking into account this result, and the fact that η is an arbitrary process, we deduce the desired stochastic fractional Euler–Lagrange equation: $\partial_2 L + {}_t^s D_b^\alpha[\partial_3 L] + {}_a^s D_t^\alpha[\partial_4 L] = 0$. The proof is complete. □

By adopting the same method as in the proof of Theorem 1 and using our result of integration by parts for stochastic Caputo fractional derivatives, i.e., item (iii) of Lemma 1, we obtain the appropriate stochastic Caputo fractional Euler–Lagrange necessary optimality condition.

Theorem 2 (The stochastic Caputo fractional Euler–Lagrange equation). *If $J \in C^1(H \times \mathbb{R} \times \mathbb{R} \to \mathbb{R})$ and $X \in C^1(I \to H)$ is an F-adapted stochastic process on $[a,b]$ with $E(X(t)) \in AC([a,b])$ that is a minimizer of*

$$J[X] = E\left(\int_a^b L\left(t, X(t), {}_a^{sC} D_t^\alpha X(t), {}_t^{sC} D_b^\alpha X(t)\right) dt\right)$$

subject to the fixed end points $E(X(a)) = X_a$ and $E(X(b)) = X_b$, then X satisfies the following stochastic fractional Euler–Lagrange equation:

$$\frac{\partial L}{\partial X} + {}_t^{sC} D_b^\alpha \left[\frac{\partial L}{\partial {}_a^{sC} D_t^\alpha}\right] + {}_a^{sC} D_t^\alpha \left[\frac{\partial L}{\partial {}_t^{sC} D_b^\alpha}\right] = 0.$$

Remark 2. *Note that the conclusions of Theorems 1 and 2 are not contradictory: one conclusion is valid for Riemann–Liouville derivative problems, while the other holds true for Caputo-type problems. The conclusions are proved in a similar manner by remarking that the additional quantity with parentheses, in the integration by parts theorem linked to the Caputo approach, vanishes under the condition that X and X^* verify the same initial and final conditions. Note also that the assumptions of Theorems 1 and 2 are necessary for the existence of left and right stochastic Riemann–Liouville/Caputo fractional derivative operators.*

Our Theorems 1 and 2 give an extension of the Euler–Lagrange equations of the classical calculus of variations [24], stochastic calculus of variations [1], and fractional calculus of variations [5].

5. Examples

The best way to illustrate a new theory is by choosing simple examples. We give two illustrative examples of the stochastic Riemann–Liouville fractional Euler–Lagrange equation proved in Section 4: the first one inspired from quantum mechanics; the second chosen to allow a simple numerical solution to the obtained stochastic Riemann–Liouville fractional Euler–Lagrange equation.

Example 1. *Let us consider the stochastic fractional variational problem (1)–(2) with*

$$L\left(t, X(t), {}_a^s D_t^\alpha X(t), {}_t^s D_b^\alpha X(t)\right) = \frac{1}{2}\left(\frac{1}{2}m \mid {}_a^s D_t^\alpha X(t) \mid^2 + \frac{1}{2}m \mid {}_t^s D_b^\alpha X(t) \mid^2\right) - V(X(t)),$$

where X is a stochastic process of second order with $E(X(t)) \in AC([a,b])$ and V maps $C'(I \to H)$ to $\mathbb{R}$. Note that

$$\frac{1}{2}\left(\frac{1}{2}m \mid {}_a^sD_t^\alpha X(t) \mid^2 + \frac{1}{2}m \mid {}_t^sD_b^\alpha X(t) \mid^2\right)$$

can be viewed as a generalized kinetic energy in the quantum mechanics framework. By applying our Theorem 1 to the current variational problem, we get

$$\frac{1}{2}m\left[{}_a^sD_t^\alpha({}_t^sD_b^\alpha X(t)) + {}_t^sD_b^\alpha({}_a^sD_t^\alpha X(t))\right] = \operatorname{grad} V(X(t)), \tag{3}$$

where gradV is the gradient of V, which in this case means the derivative of the potential energy of the system. We observe that if α tends to zero and X is a deterministic function, then relation (3) becomes what is known in the physics literature as Newton's dynamical law: $m\ddot{X}(t) = \operatorname{grad} V(X(t))$.

The calculus of variations can assist us both analytically and numerically. Now we give a numerical example, carried out with the help of the MATLAB computing environment [25].

Example 2. Let $\alpha := 0.25$, $a := 0.01$, $b := 0.99$, $X_a := 1.00$, and $X_b := 1.00$. Consider the following variational problem (1)–(2):

$$J[X] = \int_a^b {}_a^sD_t^\alpha X(t) \times {}_t^sD_b^\alpha X(t)\,dt \longrightarrow \min,$$

$$E(X(a)) = X_a, \quad E(X(b)) = X_b,$$

where $X \in C^1(I \to H)$ with $E(X(t)) \in AC$ and ${}_a^sD_t^\alpha X$ and ${}_t^sD_b^\alpha X$ denote, respectively, the left and the right stochastic fractional Riemann–Liouville derivatives of order α. Resorting again to Theorem 1, we obtain the following stochastic fractional Euler–Lagrange differential equation:

$$ {}_a^sD_t^{2\alpha}X(t) + {}_t^sD_b^{2\alpha}X(t) = 0. $$

Following [7], we observe that ${}_a^sD_t^\alpha X(t)$ and ${}_t^sD_b^\alpha X(t)$ can be approximated as follows:

$$ {}_a^sD_t^\alpha X(t) = {}_aD_t^\alpha E(X(t)) \simeq \sum_{k=0}^N \frac{(-1)^{(k-1)}\alpha (E(X(t)))^{(k)}}{k!(k-\alpha)\gamma(1-\alpha)}(t-a)^{(k-\alpha)} $$

and

$$ {}_t^sD_b^\alpha X(t) = {}_tD_b^\alpha E(X(t)) \simeq \sum_{k=0}^N \frac{-\alpha(E(X(t)))^{(k)}}{k!(k-\alpha)\gamma(1-\alpha)}(b-t)^{(k-\alpha)}. $$

Choosing $N = 1$, we get the curve for $E(X(t))$ as shown in Figure 1.

One can increase the value of N under the condition one adds a sufficient number of initial values related to some degrees of derivatives of $E(X(t))$. This particular question is similar to the standard fractional calculus and we refer the interested reader to the book [7].

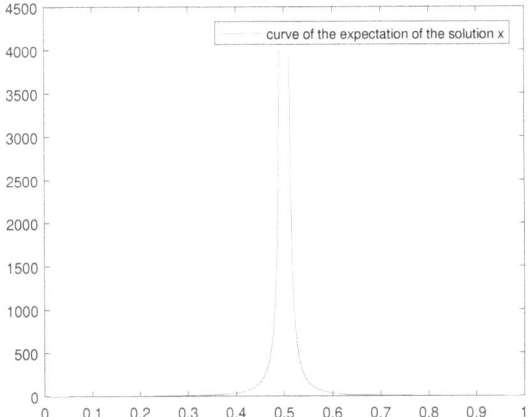

Figure 1. Expectation of the extremal to the stochastic fractional problem of the calculus of variations of Example 2.

6. Conclusions

Numerous works related to the calculus of variations, addressing different optimization problems by means of classical, stochastic, and fractional derivatives through appropriate Euler-Lagrange equations, exist in the literature. To extend available results to a stochastic-fractional framework, we have established in this work new definitions associated to left and right stochastic Riemann–Liouville/Caputo fractional integrals and derivatives, together with some properties of boundedness, linearity, additivity and interaction between involved operators. Furthermore, we have proven new integration by parts theorems, according to the novel definitions, which have a central role for the establishment of the stochastic Riemann–Liouville/Caputo fractional Euler–Lagrange equations. The obtained stochastic Riemann–Liouville/Caputo fractional Euler–Lagrange equations generalize those available on the literature of fractional calculus. Moreover, the results of the paper motivate readers and researchers to go on and further develop the theory now initiated.

It is important to note that the mathematical background used in the original fractional calculus differs from what we have established here for the stochastic fractional case. Additionally, the six constructed definitions for the left and right stochastic Riemann–Liouville/Caputo integral/derivative operators, as well as proved integration by parts formulas, differ totally to those available in the fractional calculus theory: the first are applied to second order stochastic processes, and the second act on deterministic absolute continuous functions. Furthermore, our stochastic fractional Euler–Lagrange equations serve as necessary optimality conditions to optimization problems subject to unknown stochastic processes that can be effectively approximated by numerical methods: such equations might be transformed to ones subject to unknown deterministic functions that are the expectation $E(X(t))$, for instance, when the random variables $X(t)$ follow the assumption of normality, which is instructive to approximate its expectation via stochastic fractional Euler–Lagrange equations determined statistically by the hypothesis test in inferential statistics.

We claim that the new mathematical concepts we have introduced here are more natural than those already available in the literature, since it is intuitive and convenient to proceed via application of the expectation.

Author Contributions: The authors H.Z., D.F.M.T. equally contributed to this paper, read and approved the final manuscript. All authors have read and agreed to the published version of the manuscript.

Funding: This research was funded by the Portuguese Foundation for Science and Technology (FCT), grant number UIDB/04106/2020 (CIDMA).

Acknowledgments: The authors are grateful to four anonymous reviewers for all their questions, comments and suggestions, which helped them to improve the clarity and quality of the paper.

Conflicts of Interest: The authors declare no conflict of interest.

References

1. Yasue, K. Stochastic calculus of variations. *J. Funct. Anal.* **1981**, *41*, 327–340. [CrossRef]
2. Djordjevic, J.; Silva, C.J.; Torres, D.F.M. A stochastic SICA epidemic model for HIV transmission. *Appl. Math. Lett.* **2018**, *84*, 168–175. [CrossRef]
3. Gani, S.R.; Halawar, S.V. Optimal control analysis of deterministic and stochastic epidemic model with media awareness programs. *Int. J. Optim. Control. Theor. Appl. IJOCTA* **2019**, *9*, 24–35. [CrossRef]
4. Okur, N.; Iscan, I.; Yuksek Dizdar, E. Hermite-Hadamard type inequalities for p-convex stochastic processes. *Int. J. Optim. Control. Theor. Appl. IJOCTA* **2019**, *9*, 148–153. [CrossRef]
5. Malinowska, A.B.; Torres, D.F.M. *Introduction to the Fractional Calculus of Variations*; Imperial College Press: London, UK, 2012. [CrossRef]
6. Almeida, R.; Tavares, D.; Torres, D.F.M. *The Variable-Order Fractional Calculus of Variations*; Briefs in Applied Sciences and Technology; Springer: Cham, Switzerland, 2019. [CrossRef]
7. Almeida, R.; Pooseh, S.; Torres, D.F.M. *Computational Methods in the Fractional Calculus of Variations*; Imperial College Press: London, UK, 2015. [CrossRef]
8. Baleanu, D.; Blaszczyk, T.; Asad, J.; Alipour, M. Numerical Study for Fractional Euler–Lagrange Equations of a Harmonic Oscillator on a Moving Platform. *Acta Phys. Pol. A* **2016**, *130*, 688–691. [CrossRef]
9. Baleanu, D.; Jajarmi, A.; Sajjadi, S.S.; Asad, J.H. The fractional features of a harmonic oscillator with position-dependent mass. *Commun. Theor. Phys.* **2020**, *72*, 055002. [CrossRef]
10. Ali, H.M. New approximate solutions to fractional smoking model using the generalized Mittag-Leffler function method. *Prog. Fract. Differ. Appl.* **2019**, *5*, 319–326. [CrossRef]
11. Baleanu, D.; Jajarmi, A.; Mohammadi, H.; Rezapour, S. A new study on the mathematical modelling of human liver with Caputo–Fabrizio fractional derivative. *Chaos Solitons Fractals* **2020**, *134*, 109705. [CrossRef]
12. Jajarmi, A.; Yusuf, A.; Baleanu, D.; Inc, M. A new fractional HRSV model and its optimal control: A non-singular operator approach. *Physica A* **2020**, *547*, 123860. [CrossRef]
13. Yousef, A.M.; Rida, S.Z.; Gouda, Y.G.; Zaki, A.S. On dynamics of a fractional-order SIRS epidemic model with standard incidence rate and its discretization. *Prog. Fract. Differ. Appl.* **2019**, *5*, 297–306. [CrossRef]
14. Rosa, S.; Torres, D.F.M. Optimal control and sensitivity analysis of a fractional order TB model. *Stat. Optim. Inf. Comput.* **2019**, *7*, 617–625. [CrossRef]
15. El-Sayed, A.M.A. On the stochastic fractional calculus operators. *J. Fract. Calc. Appl.* **2015**, *6*, 101–109.
16. El-Sayed, A.M.; El-Sayed, M.A.; El-Tawil, M.A.; Saif, M.S.M.; Hafiz, F.M. The mean square Riemann-Liouville stochastic fractional derivative and stochastic fractional order differential equation. *Math. Sci. Res. J.* **2005**, *9*, 142–150.
17. Hafez, F.M.; El-Sayed, A.M.A.; El-Tawil, M.A. On a stochastic fractional calculus. *Fract. Calc. Appl. Anal.* **2001**, *4*, 81–90.
18. Hafiz, F.M. The fractional calculus for some stochastic processes. *Stoch. Anal. Appl.* **2004**, *22*, 507–523. [CrossRef]
19. Samko, S.G.; Kilbas, A.A.; Marichev, O.I. *Fractional Integrals and Derivatives*; Gordon and Breach Science Publishers: Amsterdam, The Netherlands, 1993.
20. Garbaczewski, P. Fractional Laplacian and Lévy flights in bounded domains. *Acta Phys. Polon. B* **2018**, *49*, 921–942. [CrossRef]
21. Almeida, R.; Torres, D.F.M. Necessary and sufficient conditions for the fractional calculus of variations with Caputo derivatives. *Commun. Nonlinear Sci. Numer. Simul.* **2011**, *16*, 1490–1500. [CrossRef]
22. Odzijewicz, T.; Torres, D.F.M. The generalized fractional calculus of variations. *Southeast Asian Bull. Math.* **2014**, *38*, 93–117.

23. Bahaa, G.M.; Torres, D.F.M. Time-fractional optimal control of initial value problems on time scales. In *Nonlinear Analysis and Boundary Value Problems*; Springer: Cham, Switzerland, 2019; Volume 292, pp. 229–242. [CrossRef]
24. van Brunt, B. *The Calculus of Variations*; Springer: New York, NY, USA, 2004. [CrossRef]
25. Duffy, D.G. *Advanced Engineering Mathematics with MATLAB*, 4th ed.; Advances in Applied Mathematics; CRC Press: Boca Raton, FL, USA, 2017.

© 2020 by the authors. Licensee MDPI, Basel, Switzerland. This article is an open access article distributed under the terms and conditions of the Creative Commons Attribution (CC BY) license (http://creativecommons.org/licenses/by/4.0/).

Article
Modified Mittag-Leffler Functions with Applications in Complex Formulae for Fractional Calculus

Arran Fernandez [1],* and Iftikhar Husain [2]

[1] Department of Mathematics, Faculty of Arts and Sciences, Eastern Mediterranean University, via Mersin-10, Famagusta 99628, Northern Cyprus, Turkey
[2] Department of Applied Sciences and Humanities, University Polytechnic, Jamia Millia Islamia, New Delhi 110025, India; iftiphd@jmi.ac.in
* Correspondence: arran.fernandez@emu.edu.tr

Received: 9 August 2020; Accepted: 9 September 2020; Published: 12 September 2020

Abstract: Mittag-Leffler functions and their variations are a popular topic of study at the present time, mostly due to their applications in fractional calculus and fractional differential equations. Here we propose a modification of the usual Mittag-Leffler functions of one, two, or three parameters, which is ideally suited for extending certain fractional-calculus operators into the complex plane. Complex analysis has been underused in combination with fractional calculus, especially with newly developed operators like those with Mittag-Leffler kernels. Here we show the natural analytic continuations of these operators using the modified Mittag-Leffler functions defined in this paper.

Keywords: Mittag-Leffler functions; Prabhakar fractional calculus; Atangana–Baleanu fractional calculus; complex integrals; analytic continuation

MSC: 33E12; 26A33; 30B40

1. Introduction

The study of special functions has been a significant subfield of mathematical analysis for decades, connecting with other areas such as differential equations, fractional calculus, and mathematical physics [1–3]. One important class of special functions consists of the so-called Mittag-Leffler function and its extensions. These have been intensively studied, with at least one whole textbook dedicated to them [4], along with many book chapters and important research papers [5–8]. They are particularly useful due to their connections with fractional calculus, having been called "fractional exponential functions" and arising naturally in solutions to various fractional differential equations [7,9,10], including some which are useful in applications such as viscoelasticity and evolution processes [11,12].

The original Mittag-Leffler function $E_\alpha(z)$ depends on one variable z and one parameter α, and it is defined by [13]

$$E_\alpha(z) = \sum_{n=0}^{\infty} \frac{z^n}{\Gamma(n\alpha + 1)}, \qquad (1)$$

where the series is locally uniformly convergent for any $z \in \mathbb{C}$ and any $\alpha \in \mathbb{C}$ with $\mathrm{Re}(\alpha) > 0$.

This definition has been extended in various ways. The best-known extensions are the functions $E_{\alpha,\beta}(z)$ and $E_{\alpha,\beta}^\gamma(z)$, depending on one variable z and two or three parameters α, β, and γ. These are defined as follows [4,14]:

$$E_{\alpha,\beta}(z) = \sum_{n=0}^{\infty} \frac{z^n}{\Gamma(n\alpha + \beta)}, \qquad (2)$$

$$E_{\alpha,\beta}^\gamma(z) = \sum_{n=0}^{\infty} \binom{-\gamma}{n} \cdot \frac{(-z)^n}{\Gamma(n\alpha + \beta)} = \sum_{n=0}^{\infty} \frac{\Gamma(\gamma + n)}{\Gamma(\gamma)\Gamma(n\alpha + \beta)} \cdot \frac{z^n}{n!}, \qquad (3)$$

where again both series are locally uniformly convergent for any $z \in \mathbb{C}$ and any $\alpha, \beta, \gamma \in \mathbb{C}$ with $\operatorname{Re}(\alpha) > 0$. Other extensions involve even more than three parameters, or replacing the single variable z by multiple variables [10,15–17].

In fractional calculus—the study of the integral and derivative operators of calculus taken to non-integer orders [18–20]—most studies take place only in the real line. The standard Riemann–Liouville definition of a fractional integral to order α is

$$^{RL}_{a}I^{\alpha}_{x}f(x) = \frac{1}{\Gamma(\alpha)} \int_{a}^{x} (x-\xi)^{\alpha-1} f(\xi) \, d\xi,$$

where $f(x)$ is a function defined on a real interval $x \in [a,b]$ but α is permitted to be complex (with positive real part). The fractional derivative is then defined as an extension of this, by means of the following formula for $\operatorname{Re}(\alpha) \geq 0$:

$$^{RL}_{a}D^{\alpha}_{x}f(x) = \frac{d^m}{dx^m} \, ^{RL}_{a}I^{m-\alpha}_{x}f(x), \qquad m := \lfloor \operatorname{Re}(\alpha) \rfloor + 1.$$

By treating the parameter α as an independent complex variable, it can be shown that $^{RL}_{a}D^{-\alpha}_{x}f(x) = \, ^{RL}_{a}I^{\alpha}_{x}f(x)$ is an analytic extension of $^{RL}_{a}I^{\alpha}_{x}f(x)$ from the right half-plane to the left one.

For analytic complex-valued functions f, there is another formula equivalent to Riemann–Liouville which is more useful in the context of complex analysis [19,21]. Namely, the fractional differintegral (valid for all $\alpha \in \mathbb{C} \setminus \mathbb{Z}^{-}$) of $f(z)$ is

$$^{C}_{a}D^{\alpha}_{z}f(z) = \frac{\Gamma(\alpha+1)}{2i\pi} \int_{H^{z}_{a}} (\zeta - z)^{-\alpha-1} f(\zeta) \, d\zeta, \qquad (4)$$

where the complex contour of integration H^{z}_{a} is the Hankel-type contour which starts above a on the branch cut from z, wraps around z in a counterclockwise sense, and returns to a.

There are many other ways to define fractional integrals and fractional derivatives, often inspired by or related to the Riemann–Liouville definition. Some of these are discussed in [22–24], with reference to some general classes into which such operators can be classified. In pure mathematics, ideally we consider the most general possible setting in which a particular result or behaviour can be proved. In applications, of course it is necessary to consider specific types of fractional calculus for the modelling of a given real-world problem.

We have already mentioned how Mittag-Leffler functions emerge naturally from the study of fractional calculus and fractional differential equations. They also appear frequently as the kernels of fractional integral and derivative operators. Many such operators are special cases of the Prabhakar fractional calculus [14,25], which is based on the 3-parameter Mittag-Leffler function (3), and which itself can be seen as a special case of some even more general operators [17,26].

Some of these special cases were defined without realising them as special cases, and hence they were given their own names independently. Among the most intensively used types of fractional calculus in the last few years are the so-called Atangana–Baleanu operators, defined in [27] using the 1-parameter Mittag-Leffler function (1). Although integral operators using 1-parameter Mittag-Leffler kernels were already considered years earlier [28–33], the so-called AB operators have become very popular with over 350 papers published on them between 2016 and April 2020 [34]. One mathematical development in this setting has been, in [35], the extension of complex contour integral formulae like (4) to other types of fractional calculus. Doing this for AB derivatives involved the introduction of a modified Mittag-Leffler function, related but different to the original function defined by (1).

In the current work, we seek to extend the notion of this modified 1-parameter Mittag-Leffler function to define similarly modified Mittag-Leffler functions with two and three parameters. We shall perform a rigorous analysis of these modified Mittag-Leffler functions, their domains and convergence properties, and use them to extend the Atangana–Baleanu and Prabhakar fractional-calculus operators into the setting of complex variables.

Specifically, the organisation of this paper is as follows. Section 2 introduces the modified Mittag-Leffler functions, firstly re-checking the 1-parameter function in Section 2.1 and then defining the 2-parameter and 3-parameter extensions in Section 2.2. Section 3 examines how they may be used in fractional calculus, firstly for the Prabhakar operators in Section 3.1 and then for the Atangana–Baleanu operators in Section 3.2, with some further related remarks about extensions of fractional-calculus operators in Section 3.3. Finally, Section 4 concludes the paper.

2. Modified Mittag-Leffler Functions

2.1. A Rigorous Recap of the 1-Parameter Case

In this section, we re-analyse the 1-parameter modified Mittag-Leffler function defined in [35]. It is necessary to do this because there were some omissions in the work of [35]: specifically, the problems arising from the $n = 0$ term. In fact, the function cannot actually be defined in exactly the way it was in [35], because $\Gamma(-n\alpha)$ is not defined at $n = 0$. Therefore, we consider here a slightly different version which starts from $n = 1$.

Definition 1 ([35]). *The modified Mittag-Leffler function $E^\alpha(z)$ is defined by the following series for all $z \in \mathbb{C}$ and $\alpha \in \mathbb{C} \setminus \mathbb{R}$ with $\mathrm{Re}(\alpha) > 0$:*

$$E^\alpha(z) = \sum_{n=1}^{\infty} \Gamma(-n\alpha) z^n, \tag{5}$$

and by analytic continuation for all $\alpha \in \mathbb{C} \setminus \mathbb{R}$.

The reason for defining and studying this function is only to demonstrate the convergence principles and methods which will then be used for the 2-parameter and 3-parameter modified Mittag-Leffler functions in Section 2.2 below. In itself, this function may not be important, because of the missing $n = 0$ term, but we can see it as a practice "toy" case for establishing the ideas to be used later.

Of course, changing the definition of the modified 1-parameter Mittag-Leffler function will affect the results of [35] on the Atangana–Baleanu fractional derivatives. We resolve this issue in Section 3 below by finding new complex contour formulae for the Atangana–Baleanu fractional derivatives.

The following result was already proved in [35]. We reproduce the proof here, with a little more detail, and also give an alternative method of proof which will be useful later in this paper.

Proposition 1 ([35]). *The infinite power series (5) is locally uniformly convergent for all $z \in \mathbb{C}$, for any fixed $\alpha \in \mathbb{C} \setminus \mathbb{R}$ with $\mathrm{Re}(\alpha) > 0$.*

Proof using reflection formula [35]. We rewrite the power series (5) using the reflection formula for the gamma function:

$$\begin{aligned} E^\alpha(z) &= \sum_{n=1}^{\infty} \frac{\pi}{\sin(-\pi n \alpha)} \cdot \frac{1}{\Gamma(n\alpha + 1)} z^n \\ &= 2\pi i \sum_{n=1}^{\infty} \frac{1}{\exp(-i\pi n\alpha) - \exp(i\pi n\alpha)} \cdot \frac{z^n}{\Gamma(n\alpha + 1)}. \end{aligned}$$

The latter series is identical to the original Mittag-Leffler function (1) except for the extra factor $\frac{1}{\exp(-i\pi n\alpha) - \exp(i\pi n\alpha)}$. We split into two cases to consider the behaviour of this factor:

- If $\mathrm{Im}(\alpha) > 0$, then $\exp(-i\pi n\alpha) - \exp(i\pi n\alpha) \sim \exp(-i\pi n\alpha)$ and so $\frac{1}{\exp(-i\pi n\alpha) - \exp(i\pi n\alpha)} \sim \exp(i\pi n\alpha)$ has exponential decay.

- If $\text{Im}(\alpha) < 0$, then $\exp(-i\pi n\alpha) - \exp(i\pi n\alpha) \sim \exp(i\pi n\alpha)$ and so $\frac{1}{\exp(-i\pi n\alpha) - \exp(i\pi n\alpha)} \sim \exp(-i\pi n\alpha)$ has exponential decay.

Either way, for $\text{Re}(\alpha) > 0$ and $\alpha \notin \mathbb{R}$, the series converges absolutely and locally uniformly, just like the original series (1). □

Proof using ratio test and Stirling's formula. This is a more elementary way to prove convergence of a power series, going back to basics with the ratio test instead of relying on knowledge of the series for the original Mittag-Leffler function. We use Stirling's formula for the asymptotics of the gamma functions for large n; the ratio between consecutive terms is

$$\frac{a_{n+1}}{a_n} = \frac{\Gamma(-n\alpha - \alpha)}{\Gamma(-n\alpha)} z$$

$$\sim \frac{\sqrt{\frac{2\pi}{-n\alpha-\alpha}} \left(\frac{-n\alpha-\alpha}{e}\right)^{-n\alpha-\alpha}}{\sqrt{\frac{2\pi}{-n\alpha}} \left(\frac{-n\alpha}{e}\right)^{-n\alpha}} z$$

$$\sim \left(\frac{n+1}{n}\right)^{-n\alpha} \left(\frac{-\alpha}{e}\right)^{-\alpha} (n+1)^{-\alpha} z \sim \left(-\alpha(n+1)\right)^{-\alpha} z$$

as $n \to \infty$. The limit is zero if $\text{Re}(\alpha) > 0$, so in this case the series converges absolutely and locally uniformly as required. We still require the assumption $\alpha \notin \mathbb{R}$ to avoid having any zero terms. □

The following result was stated in [35], but the proof was only outlined. We present here the complete proof.

Proposition 2 ([35]). *The modified Mittag-Leffler function $E^\alpha(z)$, defined for $\text{Re}(\alpha) > 0$ by Definition 1, has an analytic continuation to all $\alpha \in \mathbb{C}\backslash\mathbb{R}$ given by the following complex integral:*

$$E^\alpha(z) = \frac{1}{-2i} \int_H e^t t^{-1} \mathfrak{S}_\alpha(zt^{-\alpha}) \, dt,$$

where H is the standard Hankel contour (starting and ending at negative real infinity and wrapping counterclockwise around the origin) and $\mathfrak{S}_\alpha$ is the function defined by

$$\mathfrak{S}_\alpha(x) = \sum_{n=1}^{\infty} \frac{x^n}{\sin(\pi n\alpha)}, \qquad x \in \mathbb{R}, \alpha \in \mathbb{C}\backslash\mathbb{R}.$$

Proof. We follow the method of [36], and proceed as follows using the standard contour integral representation of the inverse gamma function:

$$E^\alpha(z) = \sum_{n=1}^{\infty} \frac{\pi}{\sin(-\pi n\alpha)} \cdot \frac{z^n}{\Gamma(n\alpha + 1)}$$

$$= \sum_{n=1}^{\infty} \frac{\pi z^n}{\sin(-\pi n\alpha)} \cdot \frac{1}{2\pi i} \int_H e^t t^{-n\alpha - 1} \, dt$$

$$= \frac{1}{-2i} \sum_{n=1}^{\infty} \int_H e^t t^{-n\alpha - 1} \frac{z^n}{\sin(\pi n\alpha)} \, dt$$

$$= \frac{1}{-2i} \int_H e^t t^{-1} \sum_{n=1}^{\infty} \frac{(zt^{-\alpha})^n}{\sin(\pi n\alpha)} \, dt,$$

where the interchange of summation and integration is permitted by locally uniform convergence of the series. Note that locally uniform convergence of the series for $\mathfrak{S}_\alpha$ is guaranteed by the ratio test combined with an exponential-decay argument for dividing by a sine function similar to that in the first proof of Proposition 1 above. □

2.2. Extension to the 2-Parameter and 3-Parameter Cases

The original Mittag-Leffler function (1) has been modified, using the functional equation for the gamma function, as described in Definition 1 and the subsequent discussion. This modified Mittag-Leffler function $E^\alpha(z)$ depends on one variable z and one parameter α, just like the original Mittag-Leffler function $E_\alpha(z)$.

In a similar way, it is also possible to modify the 2-parameter Mittag-Leffler function (2) and the 3-parameter Mittag-Leffler function (3), thereby obtaining modified Mittag-Leffler functions with two and three parameters. We start with the 2-parameter version.

Definition 2. *The modified 2-parameter Mittag-Leffler function $E^{\alpha,\beta}(z)$ is defined by the following series for all $z \in \mathbb{C}$ and $\alpha, \beta \in \mathbb{C}$ satisfying $\operatorname{Re}(\alpha) > 0$ and α, β not both real and $n\alpha + \beta \notin \mathbb{N}$ for any $n \in \mathbb{N}$:*

$$E^{\alpha,\beta}(z) = \sum_{n=0}^{\infty} \Gamma(1 - n\alpha - \beta) z^n \quad (6)$$

and by analytic continuation for all $\alpha, \beta \in \mathbb{C}$ satisfying α, β not both real and $n\alpha + \beta \notin \mathbb{N}$ for any $n \in \mathbb{N}$.

Theorem 1. *The infinite power series (6) is locally uniformly convergent for all $z \in \mathbb{C}$, for any fixed $\alpha, \beta \in \mathbb{C}$ satisfying $\operatorname{Re}(\alpha) > 0$ and α, β not both real and $n\alpha + \beta \notin \mathbb{N}$ for any $n \in \mathbb{N}$.*

Proof using reflection formula. This follows similar lines as the first proof of Proposition 1:

$$E^{\alpha,\beta}(z) = \sum_{n=0}^{\infty} \frac{\pi}{\sin(\pi(n\alpha + \beta))} \cdot \frac{z^n}{\Gamma(n\alpha + \beta)}$$

$$= 2\pi i \sum_{n=0}^{\infty} \frac{1}{\exp(i\pi(n\alpha + \beta)) - \exp(-i\pi(n\alpha + \beta))} \cdot \frac{z^n}{\Gamma(n\alpha + \beta)},$$

where this series is identical to the original Mittag-Leffler function (2) except for the extra factor involving two exponential functions in the denominator. We split into three cases to consider the behaviour of this factor:

- If $\operatorname{Im}(\alpha) > 0$, then $\exp(i\pi(n\alpha + \beta)) - \exp(-i\pi(n\alpha + \beta)) \sim \exp(-i\pi(n\alpha + \beta))$ for sufficiently large n, and so

$$\frac{1}{\exp(i\pi(n\alpha + \beta)) - \exp(-i\pi(n\alpha + \beta))} \sim \exp(i\pi(n\alpha + \beta))$$

  has exponential decay as $n \to \infty$.

- If $\operatorname{Im}(\alpha) < 0$, then $\exp(i\pi(n\alpha + \beta)) - \exp(-i\pi(n\alpha + \beta)) \sim \exp(i\pi(n\alpha + \beta))$ for sufficiently large n, and so

$$\frac{1}{\exp(i\pi(n\alpha + \beta)) - \exp(-i\pi(n\alpha + \beta))} \sim \exp(-i\pi(n\alpha + \beta))$$

  has exponential decay as $n \to \infty$.

- If $\operatorname{Im}(\alpha) = 0$, then $\operatorname{Im}(\beta) \neq 0$ by assumption. The extra term is bounded by a constant as $n \to \infty$, namely, either

$$\frac{1}{\exp(\pi \operatorname{Im} \beta) - \exp(-\pi \operatorname{Im} \beta)} \quad \text{or} \quad \frac{1}{\exp(-\pi \operatorname{Im} \beta) - \exp(\pi \operatorname{Im} \beta)}$$

  according to the sign of $\operatorname{Im}(\beta)$.

In any case, provided $\operatorname{Re}(\alpha) > 0$ so that the original series (2) converges, the new series (6) also converges absolutely and locally uniformly. We assume throughout that the bottom never cancels out exactly to zero; i.e., that $n\alpha + \beta$ is never an integer for any n. □

Proof using ratio test and Stirling's formula. Again the calculations here are similar to those in the second proof of Proposition 1:

$$\frac{a_{n+1}}{a_n} = \frac{\Gamma(1 - (n+1)\alpha - \beta)}{\Gamma(1 - n\alpha - \beta)}$$

$$\sim \frac{\left(\frac{-(n+1)\alpha - \beta}{e}\right)^{-(n+1)\alpha - \beta} \sqrt{2\pi(-(n+1)\alpha - \beta)}}{\left(\frac{-n\alpha - \beta}{e}\right)^{-n\alpha - \beta} \sqrt{2\pi(-n\alpha - \beta)}}$$

$$\sim \left(\frac{-(n+1)\alpha - \beta}{e}\right)^{-\alpha} \left(\frac{(n+1)\alpha + \beta}{n\alpha + \beta}\right)^{-\beta} \left(\frac{(n+1)\alpha + \beta}{n\alpha + \beta}\right)^{-n\alpha}$$

$$\sim \frac{e^\alpha}{(-(n+1)\alpha - \beta)^\alpha} \left(\frac{n\alpha + \beta}{(n+1)\alpha + \beta}\right)^{n\alpha}$$

$$\sim \frac{1}{(-(n+1)\alpha - \beta)^\alpha}$$

as $n \to \infty$. The limit is zero if $\operatorname{Re}(\alpha) > 0$, so in this case the series converges absolutely and locally uniformly as required. We still require the assumption $n\alpha + \beta \notin \mathbb{N}$ to avoid having any zero terms. □

Theorem 2 (Complex integral representation of modified 2-parameter Mittag-Leffler function). *The modified 2-parameter Mittag-Leffler function, defined above under the assumption $\operatorname{Re}(\alpha) > 0$, has an analytic continuation to $\alpha, \beta \in \mathbb{C}$ satisfying α, β not both real and $n\alpha + \beta \notin \mathbb{N}$ for any $n \in \mathbb{N}$, given by the following complex integral:*

$$E^{\alpha,\beta}(z) = \frac{1}{2i} \int_H e^t t^{-\beta} \mathfrak{S}_{\alpha,\beta}(zt^{-\alpha}) \, dt,$$

where H is the standard Hankel contour (starting and ending at negative real infinity and wrapping counterclockwise around the origin) and $\mathfrak{S}_{\alpha,\beta}$ is the function defined by

$$\mathfrak{S}_{\alpha,\beta}(x) = \sum_{n=0}^{\infty} \frac{x^n}{\sin(\pi(n\alpha + \beta))}, \qquad \alpha, \beta \text{ not both real, } n\alpha + \beta \notin \mathbb{N} \,\forall n.$$

Proof. Similarly to Proposition 2, we use the contour integral representation of the inverse gamma function:

$$E^{\alpha,\beta}(z) = \sum_{n=0}^{\infty} \Gamma(1 - n\alpha - \beta) z^n$$

$$= \sum_{n=0}^{\infty} \frac{\pi}{\sin(\pi(n\alpha + \beta))} \frac{z^n}{\Gamma(n\alpha + \beta)}$$

$$= \sum_{n=0}^{\infty} \frac{\pi z^n}{\sin(\pi(n\alpha + \beta))} \cdot \frac{1}{2\pi i} \int_H e^t t^{-n\alpha - \beta} \, dt$$

$$= \frac{1}{2i} \int_H e^t t^{-\beta} \left[\sum_{n=0}^{\infty} \frac{(zt^{-\alpha})^n}{\sin(\pi(n\alpha + \beta))} \right] dt$$

$$= \frac{1}{2i} \int_H e^t t^{-\beta} \mathfrak{S}_{\alpha,\beta}(zt^{-\alpha}) \, dt,$$

as required, where the interchange of summation and integration is permitted by locally uniform convergence of the series. Note that, as before, locally uniform convergence of the series for $\mathfrak{S}_{\alpha,\beta}$ is guaranteed by the ratio test combined with an exponential-decay argument for dividing by a sine function similar to that in the first proof of Theorem 1 above. □

Definition 3. *The modified 3-parameter Mittag-Leffler function $E_\gamma^{\alpha,\beta}(z)$ is defined by the following series for all $z \in \mathbb{C}$ and $\alpha, \beta, \gamma \in \mathbb{C}$ satisfying $\operatorname{Re}(\alpha) > 0$ and α, β not both real and $n\alpha + \beta \notin \mathbb{N}$ for any $n \in \mathbb{N}$:*

$$E_\gamma^{\alpha,\beta}(z) = \sum_{n=0}^{\infty} \frac{(\gamma)_n}{n!} \Gamma(1 - n\alpha - \beta) z^n = \sum_{n=0}^{\infty} \frac{\Gamma(\gamma+n)}{\Gamma(\gamma) n!} \Gamma(1 - n\alpha - \beta) z^n \quad (7)$$

and by analytic continuation for all $\alpha, \beta, \gamma \in \mathbb{C}$ satisfying α, β not both real and $n\alpha + \beta \notin \mathbb{N}$ for any $n \in \mathbb{N}$.

Theorem 3. *The infinite power series (7) is locally uniformly convergent for all $z \in \mathbb{C}$, for any fixed $\alpha, \beta, \gamma \in \mathbb{C}$ satisfying $\operatorname{Re}(\alpha) > 0$ and α, β not both real and $n\alpha + \beta \notin \mathbb{N}$ for any $n \in \mathbb{N}$.*

Proof. This follows almost directly from the result of Theorem 1, either by the first method (reflection formula) or by the second method (ratio test).

Using the reflection formula, we find

$$E_\gamma^{\alpha,\beta}(z) = 2\pi i \sum_{n=0}^{\infty} \frac{1}{\exp(i\pi(n\alpha+\beta)) - \exp(-i\pi(n\alpha+\beta))} \cdot \frac{\Gamma(\gamma+n) z^n}{\Gamma(\gamma) \Gamma(n\alpha+\beta) n!}$$

which is identical to the original 3-parameter Mittag-Leffler series (3) except for the extra factor which is exactly the same as in Theorem 1 and therefore gives the same convergence properties under the same conditions.

Using the ratio test, we find

$$\frac{a_{n+1}}{a_n} = \frac{(\gamma+n) \Gamma(1-(n+1)\alpha-\beta)}{(n+1) \Gamma(1-n\alpha-\beta)} \sim \frac{\Gamma(1-(n+1)\alpha-\beta)}{\Gamma(1-n\alpha-\beta)},$$

which is exactly the same as in Theorem 1 and therefore gives the same convergence properties under the same conditions. □

Theorem 4 (Complex integral representation of modified 3-parameter Mittag-Leffler function). *The modified 3-parameter Mittag-Leffler function, defined above under the assumption $\operatorname{Re}(\alpha) > 0$, has an analytic continuation to $\alpha, \beta, \gamma \in \mathbb{C}$ satisfying α, β not both real and $n\alpha + \beta \notin \mathbb{N}$ for any $n \in \mathbb{N}$, given by the following complex integral:*

$$E_\gamma^{\alpha,\beta}(z) = \frac{1}{2i} \int_H e^t t^{-\beta} \mathfrak{S}_{\alpha,\beta}^\gamma(z t^{-\alpha}) \, dt,$$

where H is the standard Hankel contour (starting and ending at negative real infinity and wrapping counterclockwise around the origin) and $\mathfrak{S}_{\alpha,\beta}^\gamma$ is the function defined by

$$\mathfrak{S}_{\alpha,\beta}^\gamma(x) = \sum_{n=0}^{\infty} \frac{\Gamma(\gamma+n) x^n}{\Gamma(\gamma) n! \sin(\pi(n\alpha+\beta))}, \quad \alpha, \beta \text{ not both real, } n\alpha+\beta \notin \mathbb{N} \ \forall n.$$

Proof. Similarly to Proposition 2 and Theorem 2, we use the contour integral representation of the inverse gamma function:

$$\begin{aligned}
E_\gamma^{\alpha,\beta}(z) &= \sum_{n=0}^\infty \frac{\Gamma(\gamma+n)}{\Gamma(\gamma)n!}\Gamma(1-n\alpha-\beta)z^n \\
&= \sum_{n=0}^\infty \frac{\pi}{\sin(\pi(n\alpha+\beta))}\frac{\Gamma(\gamma+n)z^n}{\Gamma(\gamma)\Gamma(n\alpha+\beta)n!} \\
&= \sum_{n=0}^\infty \frac{\pi\Gamma(\gamma+n)z^n}{\Gamma(\gamma)n!\sin(\pi(n\alpha+\beta))}\frac{1}{2\pi i}\int_H e^t t^{-n\alpha-\beta}\,dt \\
&= \frac{1}{2i}\int_H e^t t^{-\beta}\left[\sum_{n=0}^\infty \frac{\Gamma(\gamma+n)(zt^{-\alpha})^n}{\Gamma(\gamma)n!\sin(\pi(n\alpha+\beta))}\right]dt \\
&= \frac{1}{2i}\int_H e^t t^{-\beta}\mathfrak{S}_{\alpha,\beta}^\gamma(zt^{-\alpha})\,dt,
\end{aligned}$$

where the interchange of summation and integration is permitted by locally uniform convergence of the series. Note that locally uniform convergence of the series for $\mathfrak{S}_{\alpha,\beta}^\gamma$ is guaranteed by the same property of $\mathfrak{S}_{\alpha,\beta}$, since the ratio test gives almost exactly the same expression for both. □

Remark 1. *The condition required in the above definitions and theorems, that $n\alpha + \beta \notin \mathbb{N}$ for any $n \in \mathbb{N}$, may at first seem to be very restrictive. However, this is simply the requirement that all the terms of the series itself are well-defined. If we ever have $n\alpha + \beta \in \mathbb{N}$ for some n, then $\Gamma(1-n\alpha-\beta)$ is not defined for this value of n, and so the series itself makes no sense. This condition is added simply to ensure that our definitions can actually make sense, even before convergence considerations.*

3. Extensions of Fractional Operators

3.1. Contour Integral Formulae for Prabhakar Fractional Operators

Definition 4 ([14,25,37])**.** *The Prabhakar fractional integral of a function $f \in L^1[a,b]$, with parameters $\alpha,\beta,\gamma,\delta \in \mathbb{C}$ satisfying $\mathrm{Re}(\alpha) > 0$ and $\mathrm{Re}(\beta) > 0$, is defined as*

$$\,_a^P I_x^{\alpha,\beta,\gamma,\delta} f(x) = \int_a^x (x-\xi)^{\beta-1} E_{\alpha,\beta}^\gamma(\delta(x-\xi)^\alpha)f(\xi)\,d\xi, \tag{8}$$

using the 3-parameter Mittag-Leffler function (3) as a kernel function. This operator can also be written as an infinite series of Riemann–Liouville fractional integrals, as follows:

$$\,_a^P I_x^{\alpha,\beta,\gamma,\delta} f(x) = \sum_{n=0}^\infty \frac{(\gamma)_n \delta^n}{n!}\,_a^{RL} I_x^{n\alpha+\beta} f(x). \tag{9}$$

The Prabhakar fractional derivative of a smooth function $f(x)$, with parameters $\alpha,\beta,\gamma,\delta \in \mathbb{C}$ satisfying $\mathrm{Re}(\alpha) > 0$ and $\mathrm{Re}(\beta) \geq 0$, is defined as

$$\,_a^P D_x^{\alpha,\beta,\gamma,\delta} f(x) = \frac{d^m}{dx^m}\,_a^P I_x^{\alpha,m-\beta,-\gamma,\delta} f(x), \qquad m := \lfloor \mathrm{Re}(\beta) \rfloor + 1. \tag{10}$$

Using composition properties of Riemann–Liouville derivatives and integrals, this operator can be written as an infinite series similar to (9):

$$\,_a^P D_x^{\alpha,\beta,\gamma,\delta} f(x) = \sum_{n=0}^\infty \frac{(-\gamma)_n \delta^n}{n!}\,_a^{RL} I_x^{n\alpha-\beta} f(x), \tag{11}$$

where the operator denoted by ${}^{RL}_a I_x^{n\alpha-\beta}$ is either a Riemann–Liouville integral or a Riemann–Liouville derivative depending on the sign of $\operatorname{Re}(n\alpha - \beta)$.

Note that the variable x in the above definition is assumed to be real, in the fixed interval $[a,b]$. In the previous paper [35], the Atangana–Baleanu fractional operators were extended from the real line to the complex plane, using a complex contour integral approach and the modified 1-parameter Mittag-Leffler function (5). Now we seek to do the same for the Prabhakar fractional operators, using the modified 3-parameter Mittag-Leffler function which we have defined in this paper.

Theorem 5. *The analytic continuation of the Prabhakar fractional integral is given by*

$$ {}^P_a I_z^{\alpha,\beta,\gamma,\delta} f(z) = \frac{1}{2\pi i} \int_{H_a^z} (\zeta-z)^{\beta-1} E_\gamma^{\alpha,\beta}\big(\delta(\zeta-z)^\alpha\big) f(\zeta)\,\mathrm{d}\zeta, \qquad (12) $$

where $E_\gamma^{\alpha,\beta}(x)$ is the modified 3-parameter Mittag-Leffler function defined by (7) above, and the complex contour of integration H_a^z is the Hankel-type contour which starts above a on the branch cut from z, wraps around z in a counterclockwise sense, and returns to a.

This formula (12) also covers Prabhakar fractional differentiation, under the convention (following from the semigroup property and series formula) that ${}^P_a D_x^{\alpha,\beta,\gamma,\delta} f(x) = {}^P_a I_x^{\alpha,-\beta,\gamma,\delta} f(x)$. In other words, we have

$$ {}^P_a D_z^{\alpha,\beta,\gamma,\delta} f(z) = \frac{1}{2\pi i} \int_{H_a^z} (\zeta-z)^{\beta-1} E_{-\gamma}^{\alpha,-\beta}\big(\delta(\zeta-z)^\alpha\big) f(\zeta)\,\mathrm{d}\zeta. $$

The assumption on the parameters $\alpha,\beta,\gamma,\delta$ for this Theorem is that α,β not both real and $n\alpha+\beta \notin \mathbb{N}$ for any $n \in \mathbb{N}$.

Proof. We use the series formula for Prabhakar fractional calculus, noting that both (9) for integrals and (11) for derivatives become the same formula under the convention ${}^P_a D_x^{\alpha,\beta,\gamma,\delta} f(x) = {}^P_a I_x^{\alpha,-\beta,-\gamma,\delta} f(x)$. We have

$$ \begin{aligned} {}^P_a I_x^{\alpha,\beta,\gamma,\delta} f(x) &= \sum_{n=0}^\infty \frac{(\gamma)_n \delta^n}{n!} \,{}^{RL}_a I_x^{n\alpha+\beta} f(x) \\ &= \sum_{n=0}^\infty \frac{(\gamma)_n \delta^n}{n!} \,{}^{C}_a I_x^{n\alpha+\beta} f(x) \\ &= \sum_{n=0}^\infty \frac{(\gamma)_n \delta^n}{n!} \frac{\Gamma(1-n\alpha-\beta)}{2\pi i} \int_{H_a^z} (\zeta-z)^{n\alpha+\beta-1} f(\zeta)\,\mathrm{d}\zeta \\ &= \frac{1}{2\pi i} \int_{H_a^z} (\zeta-z)^{\beta-1} \left[\sum_{n=0}^\infty \frac{(\gamma)_n}{n!} \Gamma(1-n\alpha-\beta)\big(\delta(\zeta-z)^\alpha\big)^n \right] f(\zeta)\,\mathrm{d}\zeta \\ &= \frac{1}{2\pi i} \int_{H_a^z} (\zeta-z)^{\beta-1} E_\gamma^{\alpha,\beta}\big(\delta(\zeta-z)^\alpha\big) f(\zeta)\,\mathrm{d}\zeta. \end{aligned} $$

The above manipulation is valid provided that $\operatorname{Re}(\alpha) > 0$. Note that we do not need any assumption on $\operatorname{Re}(\beta)$ since the case $\operatorname{Re}(\beta) > 0$ is covered by the Prabhakar fractional integral and the case $\operatorname{Re}(\beta) \leq 0$ by the Prabhakar fractional derivative.

The final formula, however—the right-hand side of Equation (12)—is well-defined and analytic for any $\alpha,\beta,\gamma,\delta$ satisfying α,β not both real and $n\alpha+\beta \notin \mathbb{N}$ for any $n \in \mathbb{N}$, by the analytic continuation of $E_\gamma^{\alpha,\beta}$ given in Theorem 4. Therefore, (12) provides the analytic continuation of the Prabhakar fractional integral and derivative, even to the cases where $\operatorname{Re}(\alpha) > 0$ no longer applies. □

3.2. Contour Integral Formulae for Atangana–Baleanu Fractional Operators

Definition 5 ([27,38]). *The Atangana–Baleanu fractional integral of a function $f \in L^1[a,b]$, with parameter $\alpha \in (0,1)$, is defined as*

$$^{AB}_{a}I^{\alpha}_{x}f(x) = \frac{1-\alpha}{B(\alpha)}f(x) + \frac{\alpha}{B(\alpha)}\,^{RL}_{a}I^{\alpha}_{x}f(x), \tag{13}$$

The Atangana–Baleanu fractional derivatives of a function $f \in C^1[a,b]$, with parameter $\alpha \in (0,1)$, of Riemann–Liouville and Caputo types respectively, are defined as:

$$^{ABR}_{a}D^{\alpha}_{x}f(x) = \frac{B(\alpha)}{1-\alpha} \cdot \frac{d}{dx} \int_{a}^{x} E_{\alpha}\left(\frac{-\alpha}{1-\alpha}(x-\xi)^{\alpha}\right) f(\xi)\,d\xi, \tag{14}$$

$$^{ABC}_{a}D^{\alpha}_{x}f(x) = \frac{B(\alpha)}{1-\alpha} \int_{a}^{x} E_{\alpha}\left(\frac{-\alpha}{1-\alpha}(x-\xi)^{\alpha}\right) f'(\xi)\,d\xi, \tag{15}$$

using the 1-parameter Mittag-Leffler function (1) as a kernel function. These operators can also be written as infinite series of Riemann–Liouville fractional integrals, as follows:

$$^{ABR}_{a}D^{\alpha}_{x}f(x) = \frac{B(\alpha)}{1-\alpha} \cdot \frac{d}{dx} \sum_{n=0}^{\infty}\left(\frac{-\alpha}{1-\alpha}\right)^{n}\,^{RL}_{a}I^{n\alpha+1}_{x}f(x) \tag{16}$$

$$= \frac{B(\alpha)}{1-\alpha} \sum_{n=0}^{\infty}\left(\frac{-\alpha}{1-\alpha}\right)^{n}\,^{RL}_{a}I^{n\alpha}_{x}f(x), \tag{17}$$

$$^{ABC}_{a}D^{\alpha}_{x}f(x) = \frac{B(\alpha)}{1-\alpha} \sum_{n=0}^{\infty}\left(\frac{-\alpha}{1-\alpha}\right)^{n}\,^{RL}_{a}I^{n\alpha+1}_{x}f'(x). \tag{18}$$

In the paper [35], these definitions were extended beyond $\alpha \in (0,1)$ to complex values of α. The Atangana–Baleanu (AB) integral is easy to extend to complex α and complex x, just using the well-known extension of the Riemann–Liouville integral:

$$^{AB}_{a}I^{\alpha}_{z}f(z) = \frac{1}{2\pi i B(\alpha)} \int_{H^{z}_{a}} \left(\frac{1-\alpha}{\zeta - z} + \frac{\alpha \Gamma(1-\alpha)}{(\zeta-z)^{1-\alpha}}\right) f(\zeta)\,d\zeta.$$

For the AB derivatives (of both types), the complex contour formulae written in [35] were as follows:

$$^{ABR}_{a}D^{\alpha}_{z}f(z) = \frac{B(\alpha)}{2\pi i(1-\alpha)} \cdot \frac{d}{dz} \int_{H^{z}_{a}} E^{\alpha}\left(\frac{-\alpha}{1-\alpha}\zeta - z)^{\alpha}\right) f(\zeta)\,d\zeta,$$

$$^{ABC}_{a}D^{\alpha}_{z}f(z) = \frac{B(\alpha)}{2\pi i(1-\alpha)} \int_{H^{z}_{a}} E^{\alpha}\left(\frac{-\alpha}{1-\alpha}\zeta - z)^{\alpha}\right) f'(\zeta)\,d\zeta,$$

where here the notation E^{α} refers to the incorrectly defined function from [35],

$$E^{\alpha}(z) = \sum_{n=0}^{\infty} \Gamma(-n\alpha) z^{n},$$

this being incorrect because of the $n = 0$ term.

The correct version of these formulae is given by slightly modifying them as follows:

$$^{ABR}_{a}D^{\alpha}_{z}f(z) = \frac{B(\alpha)}{2\pi i(1-\alpha)} \cdot \frac{d}{dz} \int_{H^{z}_{a}} \left[\frac{1}{\zeta - z} + E^{\alpha}\left(\frac{-\alpha}{1-\alpha}\zeta - z)^{\alpha}\right)\right] f(\zeta)\,d\zeta,$$

$$^{ABC}_{a}D^{\alpha}_{z}f(z) = \frac{B(\alpha)}{2\pi i(1-\alpha)} \int_{H^{z}_{a}} \left[\frac{1}{\zeta - z} + E^{\alpha}\left(\frac{-\alpha}{1-\alpha}\zeta - z)^{\alpha}\right)\right] f'(\zeta)\,d\zeta - f(a),$$

where this time the notation E^{α} refers to the well-defined function from (5) above. These formulae are obtained by treating the $n = 0$ term separately, starting from the series formulae (16) and (18), and in the Caputo case using the fact that $^{RL}_aI^1_xf'(x) = f(x) - f(a)$. However, we can obtain more elegant formulae by considering instead the series formula (17) and using the 2-parameter modified Mittag-Leffler function defined in Section 2.2 above.

Theorem 6. *The analytic continuation of the AB fractional derivative of Riemann–Liouville type is given by*

$$^{ABR}_aD^\alpha_z f(z) = \frac{B(\alpha)}{2\pi i(1-\alpha)} \int_{H^z_a} E^{\alpha,0}\left(\frac{-\alpha}{1-\alpha}(\zeta-z)^\alpha\right) \frac{f(\zeta)}{\zeta-z} \,d\zeta, \qquad (19)$$

where $E^{\alpha,\beta}(x)$ is the modified 2-parameter Mittag-Leffler function defined by (6) above, and the complex contour of integration H^z_a is the Hankel-type contour which starts above a on the branch cut from z, wraps around z in a counterclockwise sense, and returns to a.

The analytic continuation of the AB fractional derivative of Caputo type can then be deduced using the relationship between ABR and ABC derivatives given by the fundamental theorem of calculus:

$$^{ABC}_aD^\alpha_z f(z) = \frac{B(\alpha)}{2\pi i(1-\alpha)} \int_{H^z_a} E^{\alpha,0}\left(\frac{-\alpha}{1-\alpha}(\zeta-z)^\alpha\right) \frac{f(\zeta)}{\zeta-z} \,d\zeta - \frac{B(\alpha)}{1-\alpha} E_\alpha\left(\frac{-\alpha}{1-\alpha}z^\alpha\right) f(a).$$

The assumption on the parameter α for this Theorem is simply $\alpha \in \mathbb{C}\backslash\mathbb{R}$.

Proof. We start from the series formula (17) for the ABR fractional derivative, and use the complex integral representation for the Riemann–Liouville fractional integral:

$$^{ABR}_aD^\alpha_x f(x) = \frac{B(\alpha)}{1-\alpha} \sum_{n=0}^\infty \left(\frac{-\alpha}{1-\alpha}\right)^n {}^{RL}_aI^{n\alpha}_x f(x)$$

$$= \frac{B(\alpha)}{1-\alpha} \sum_{n=0}^\infty \left(\frac{-\alpha}{1-\alpha}\right)^n \frac{\Gamma(-n\alpha+1)}{2i\pi} \int_{H^z_a} (\zeta-z)^{n\alpha-1} f(\zeta)\,d\zeta.$$

Note that this substitution is valid for all values of n, since the complex contour formula (4) is valid for any order of differentiation not in $\mathbb{Z}^-$. Here the orders of differentiation are $-n\alpha$ for $n \geq 0$, which is either zero or nonreal. (This is why, when using (16) instead of (17), we needed to treat $n=0$ separately: because in the case of (16), the $n=0$ term gives order of differentation -1 which is in $\mathbb{Z}^-$).

Continuing, and using the fact that the series formulae for AB derivatives are locally uniformly convergent [38]:

$$^{ABR}_aD^\alpha_x f(x) = \frac{B(\alpha)}{1-\alpha} \frac{1}{2i\pi} \int_{H^z_a} f(\zeta) \sum_{n=0}^\infty \Gamma(-n\alpha+1) \left(\frac{-\alpha}{1-\alpha}\right)^n (\zeta-z)^{n\alpha-1} \,d\zeta$$

$$= \frac{B(\alpha)}{1-\alpha} \frac{1}{2i\pi} \int_{H^z_a} \frac{f(\zeta)}{\zeta-z} E^{\alpha,0}\left(\frac{-\alpha}{1-\alpha}(\zeta-z)^\alpha\right) d\zeta,$$

which is the desired result for ABR derivatives.

For the case of ABC derivatives, we simply use the following relationship between ABR and ABC following from the series formulae (16)–(18):

$$\begin{aligned}
{}^{ABC}_aD^\alpha_x f(x) &= \frac{B(\alpha)}{1-\alpha} \sum_{n=0}^\infty \left(\frac{-\alpha}{1-\alpha}\right)^n {}^{RL}_aI^{n\alpha}_x \left({}^{RL}_aI^1_x f'(x)\right) \\
&= \frac{B(\alpha)}{1-\alpha} \sum_{n=0}^\infty \left(\frac{-\alpha}{1-\alpha}\right)^n {}^{RL}_aI^{n\alpha}_x \left(f(x)-f(a)\right) \\
&= {}^{ABR}_aD^\alpha_x f(x) - \frac{B(\alpha)}{1-\alpha} \sum_{n=0}^\infty \left(\frac{-\alpha}{1-\alpha}\right)^n {}^{RL}_aI^{n\alpha}_x(1)f(a) \\
&= {}^{ABR}_aD^\alpha_x f(x) - \frac{B(\alpha)}{1-\alpha} \sum_{n=0}^\infty \left(\frac{-\alpha}{1-\alpha}\right)^n \frac{x^{n\alpha}}{\Gamma(n\alpha+1)} f(a) \\
&= {}^{ABR}_aD^\alpha_x f(x) - \frac{B(\alpha)}{1-\alpha} E_\alpha\left(\frac{-\alpha}{1-\alpha}x^\alpha\right) f(a).
\end{aligned}$$

The last term here is the initial value term which gives the desired result for ABC derivatives. □

Remark 2. *Note that using $\beta = 0$ in the usual 2-parameter or 3-parameter Mittag-Leffler functions would not be possible when using them as kernel functions, because it would lead to a non-integrable singularity. However, in the complex setting, this is fine since $\frac{1}{\zeta-z}$ can be integrated using Cauchy's integral formula.*

Remark 3. *It is known that the Atangana–Baleanu fractional operators are special cases of the Prabhakar fractional calculus. Indeed, this is an obvious fact for the AB derivative, since it (like the Prabhakar operators) is defined using an integral transform with Mittag-Leffler kernel. The AB integral, on the other hand, is simply the linear combination of a function with its Riemann–Liouville integral, with no Mittag-Leffler functions involved in the definition; it was only noticed recently in [39] that it too is a special case of Prabhakar. The relationships are given by*

$$\begin{aligned}
{}^{AB}_aI^\alpha_x f(x) &= \frac{1-\alpha}{B(\alpha)}\, {}^P_aI^{\alpha,0,-1,\frac{-\alpha}{1-\alpha}}_x f(x), \\
{}^{ABR}_aD^\alpha_x f(x) &= \frac{1-\alpha}{B(\alpha)}\, {}^P_aD^{\alpha,0,-1,\frac{-\alpha}{1-\alpha}}_x f(x).
\end{aligned}$$

Using this, it is possible to deduce the result of Theorem 6 directly from that of Theorem 5. Note that the multiplier $(\zeta-z)^{\beta-1}$ appearing in (12), which is a typical power function multiplier found when dealing with Mittag-Leffler function kernels, in the AB case becomes simply $\frac{1}{\zeta-z}$, which is a typical multiplier found in complex analysis according to Cauchy's integral formula.

3.3. Series for Negative α

In the paper [36], a series formula is given for the Mittag-Leffler function $E_\alpha(z)$ which is valid for negative real numbers α, by using a functional equation that emerges from the complex integral representation. The same functional equation approach works to prove similar series formulae for the two-parameter Mittag-Leffler function $E_{\alpha,\beta}(z)$ and for complex α with $\text{Re}(\alpha) < 0$. We state the general result as follows.

Proposition 3 ([36])**.** *The analytic continuation of the two-parameter Mittag-Leffler function $E_{\alpha,\beta}(z)$, originally defined by (2), to the domain $\alpha, \beta \in \mathbb{C}, \text{Re}(\alpha) < 0$ is given by the following locally uniformly convergent series:*

$$E_{\alpha,\beta}(z) = \sum_{n=1}^\infty \frac{-z^{-n}}{\Gamma(-n\alpha+\beta)}. \tag{20}$$

Proof. From the complex integral representation of the two-parameter Mittag-Leffler function,

$$E_{\alpha,\beta}(z) = \frac{1}{2\pi i} \int_H \frac{t^{\alpha-\beta} e^t}{t^\alpha - z} \, dt = \frac{1}{2\pi i} \int_H \frac{e^t}{t^\beta - z t^{\beta-\alpha}} \, dt,$$

valid for all $\alpha, \beta \in \mathbb{C}$, we use the algebraic identity

$$\frac{1}{t^\beta - z t^{\beta-\alpha}} = \frac{1}{t^\beta} - \frac{1}{t^\beta - z^{-1} t^{\alpha+\beta}} \tag{21}$$

to obtain

$$E_{\alpha,\beta}(z) = \frac{1}{2\pi i} \int_H \frac{e^t}{t^\beta} \, dt - \frac{1}{2\pi i} \int_H \frac{e^t}{t^\beta - z^{-1} t^{\alpha+\beta}} \, dt = \frac{1}{\Gamma(\beta)} - E_{-\alpha,\beta}(z^{-1}),$$

valid for all $\alpha, \beta \in \mathbb{C}$. Then, if $\operatorname{Re}(\alpha) < 0$, we have $\operatorname{Re}(-\alpha) > 0$ and therefore we can use the original series formula (2) for $E_{-\alpha,\beta}(z^{-1})$. Cancelling the $n = 0$ term, this gives the desired series for $E_{\alpha,\beta}(z)$ in the case of $\operatorname{Re}(\alpha) < 0$. □

Remark 4. *The same technique cannot be used to give an elegant series representation of the 3-parameter Mittag-Leffler function $E_{\alpha,\beta}^\gamma(z)$ for negative values of α. This is because the complex integral representation of this function involves a γth power:*

$$E_{\alpha,\beta}^\gamma(z) = \frac{1}{2\pi i} \int_H \frac{t^{-\beta} e^t}{(1 - z t^{-\alpha})^\gamma} \, dt,$$

and there is no analogue of the identity (21) for reciprocals of γth powers.

Remark 5. *Furthermore, the technique of Proposition 3 cannot be applied to our modified Mittag-Leffler functions either, even in the 1-parameter and 2-parameter cases. The complex integral representations of Proposition 2 and Theorem 2 have integrands involving the functions $\mathfrak{S}_\alpha$ and $\mathfrak{S}_{\alpha,\beta}$ which do not have simple identities like (21) between them.*

4. Conclusions and Further Work

This paper serves as a continuation of the work of [35], in which the first modified Mittag-Leffler function was defined and used to extend Atangana–Baleanu fractional operators into the complex context. Here we have corrected an omission in [35], in which the issues surrounding the $n = 0$ term of the Mittag-Leffler series were overlooked. We have defined modified Mittag-Leffler functions of one, two, and three parameters, and rigorously checked the convergence issues for the series in each case.

The power of Mittag-Leffler functions and their series in fractional calculus cannot be understated. Several important operators of fractional calculus are defined using Mittag-Leffler functions, and the series formulae for these operators have been useful in proving a number of useful properties. Our modified Mittag-Leffler functions and their series can be used to provide new formulae for the same operators, which are valid in larger domains than the original ones. We showed how both the Prabhakar and the Atangana–Baleanu operators can be applied to find fractional derivatives and integrals of functions of a complex variable as well as real functions.

The work contained in this paper will be useful for ongoing research into these fractional-calculus operators and their applications. It was already seen, for example, in [12,40,41], that complex integral representations of Mittag-Leffler functions are useful in finding asymptotic expansions, and therefore in bounding and approximating the functions. In some cases, complex orders of fractional derivatives can be vital for modelling [42–44]. The analysis of fractional evolution processes in [12] even used Mittag-Leffler-type infinite series involving the gamma function at negative parameters, such as $\Gamma(1 - n\alpha)$, similarly to the functions we have introduced in this paper. Therefore, we expect our formulae to find applications in the future.

Furthermore, we investigated another way of extending Mittag-Leffler functions by analytic continuation, namely the series for negative α. Although this approach does not work directly on the modified Mittag-Leffler functions defined here, we believe it will be useful in the analysis of Atangana–Baleanu and Prabhakar fractional operators. Further research in this direction is currently ongoing.

Author Contributions: All authors contributed equally to this article. All authors have read and agreed to the published version of the manuscript.

Funding: This research received no external funding.

Conflicts of Interest: The authors declare no conflict of interest.

References

1. Andrews, L.C. *Special Functions of Mathematics for Engineers*; SPIE Press: Washington, DC, USA, 1998.
2. Lebedev, N.N.; Silverman, R.A. *Special Functions and Their Applications*, 2nd ed.; Dover: New York, NY, USA, 1972.
3. Nikiforov, A.F.; Uvarov, V.B. *Special Functions of Mathematical Physics: A Unified Introduction with Applications*; Birkhauser: Boston, MA, USA, 1988.
4. Gorenflo, R.; Kilbas, A.A.; Mainardi, F.; Rogosin, S.V. *Mittag-Leffler Functions, Related Topics and Applications*; Springer: Berlin, Germany, 2016.
5. Haubold, H.J.; Mathai, A.M.; Saxena, R.K. Mittag-Leffler functions and their applications. *J. Appl. Math.* **2011**, *2011*. [CrossRef]
6. Kilbas, A.A.; Saigo, M. On Mittag-Leffler type function, fractional calculas operators and solutions of integral equations. *Integral Transforms Spec. Funct.* **1996**, *4*, 355–370. [CrossRef]
7. Mathai, A.M.; Haubold, H.J. Mittag-Leffler functions and fractional calculus. In *Special Functions for Applied Scientists*; Mathai, A.M., Haubold, H.J., Eds.; Springer: Berlin, Germany, 2008; pp. 79–134.
8. Pillai, R.N. On Mittag-Leffler functions and related distributions. *Ann. Inst. Stat. Math.* **1990**, *42*, 157–161. [CrossRef]
9. Camargo, R.F.; de Oliveira, E.C.; Vaz, J. On the generalized Mittag-Leffler function and its application in a fractional telegraph equation. *Math. Phys. Anal. Geom.* **2012**, *15*, 1–16. [CrossRef]
10. Kilbas, A.A.; Srivastava, H.M.; Trujillo, J.J. *Theory and Applications of Fractional Differential Equations*; Elsevier: Amsterdam, The Netherlands, 2006.
11. Koeller, R.C. Applications of Fractional Calculus to the Theory of Viscoelasticity. *J. Appl. Mech.* **1984**, *51*, 299–307. [CrossRef]
12. Mainardi, F.; Gorenflo, R. On Mittag-Leffler-type functions in fractional evolution processes. *J. Comput. Appl. Math.* **2000**, *118*, 283–299. [CrossRef]
13. Mittag-Leffler, M.G. Sur la nouvelle fonction $E(x)$. *Comptes Rendus l'Académie Sci.* **1903**, *137*, 554–558.
14. Prabhakar, T.R. A singular integral equation with a generalized Mittag Leffler function in the kernel. *Yokohama Math. J.* **1971**, *19*, 7–15.
15. Luchko, Y. Operational method in fractional calculus. *Fract. Calc. Appl. Anal.* **1999**, *2*, 463–488.
16. Saxena, R.K.; Kalla, S.L.; Saxena, R. Multivariate analogue of generalised Mittag-Leffler function. *Integral Transforms Spec. Funct.* **2011**, *22*, 533–548. [CrossRef]
17. Srivastava, H.M.; Tomovski, Ž. Fractional calculus with an integral operator containing a generalized Mittag–Leffler function in the kernel. *Appl. Math. Comput.* **2009**, *211*, 198–210. [CrossRef]
18. Miller, K.S.; Ross, B. *An Introduction to the Fractional Calculus and Fractional Differential Equations*; Wiley: New York, NY, USA, 1993.
19. Oldham, K.B.; Spanier, J. *The Fractional Calculus*; Academic Press: San Diego, CA, USA, 1974.
20. Samko, S.G.; Kilbas, A.A.; Marichev, O.I. *Fractional Integrals and Derivatives: Theory and Applications*; Gordon & Breach Science Publishers: Yverdon, Switzerland, 1993.
21. Nekrassov, P. Sur la différentiation générale". *Matemat. Sbornik* **1888**, *14*, 45–166. (In Russian)
22. Baleanu, D.; Fernandez, A. On fractional operators and their classifications *Mathematics* **2019**, *7*, 830. [CrossRef]
23. Hilfer, R.; Luchko, Y. Desiderata for Fractional Derivatives and Integrals. *Mathematics* **2019**, *7*, 149. [CrossRef]

24. Teodoro, G.S.; Machado, J.A.T.; de Oliveira, E.C. A review of definitions of fractional derivatives and other operators. *J. Comput. Phys.* **2019**, *388*, 195–208. [CrossRef]
25. Kilbas, A.A.; Saigo, M.; Saxena, R.K. Generalized Mittag-Leffler function and generalized fractional calculus operators. *Integral Transforms Spec. Funct.* **2004**, *15*, 31–49. [CrossRef]
26. Kiryakova, V.S. Multiple (multiindex) Mittag–Leffler functions and relations to generalized fractional calculus. *J. Comput. Appl. Math.* **2000**, *118*, 241–259. [CrossRef]
27. Atangana, A.; Baleanu, D. New fractional derivative with non-local and non-singular kernel. *Ther. Sci.* **2016**, *20*, 757–763.
28. Camargo, R.F.; Chiacchio, A.O.; Charnet, R.; de Oliveira, E.C. Solution of the fractional Langevin equation and the Mittag–Leffler functions. *J. Math. Phys.* **2009**, *50*, 063507. [CrossRef]
29. Desposito, M.A.; Viñales, A.D. Memory effects in the asymptotic diffusive behavior of a classical oscillator described by a generalized Langevin equation. *Phys. Rev. E* **2008**, *77*, 031123. [CrossRef]
30. Mankin, R.; Laas, K.; Sauga, A. Generalized Langevin equation with multiplicative noise: Temporal behavior of the autocorrelation functions. *Phys. Rev. E* **2011**, *83*, 061131. [CrossRef]
31. Viñales, A.D.; Despósito, M.A. Anomalous diffusion induced by a Mittag-Leffler correlated noise. *Phys. Rev. E* **2007**, *75*, 042102. [CrossRef] [PubMed]
32. Viñales, A.D.; Paissan, G.H. Velocity autocorrelation of a free particle driven by a Mittag-Leffler noise: Fractional dynamics and temporal behaviors. *Phys. Rev. E* **2014**, *90*, 062103. [CrossRef] [PubMed]
33. Viñales, A.D.; Wang, K.G.; Desposito, M.A. Anomalous diffusive behavior of a harmonic oscillator driven by a Mittag-Leffler noise. *Phys. Rev. E* **2009**, *80*, 011101. [CrossRef]
34. Templeton, A. A bibliometric analysis of Atangana-Baleanu operators in fractional calculus. *Alex. Eng. J.* **2020**, *59*, 2733–2738. [CrossRef]
35. Fernandez, A. A complex analysis approach to Atangana–Baleanu fractional calculus. *Math. Methods Appl. Sci.* **2019**, 1–18. [CrossRef]
36. Hanneken, J.W.; Achar, B.N.N.; Puzio, R.; Vaught, D.M. Properties of the Mittag-Leffler function for negative alpha. *Phys. Scrip.* **2009**, *T136*, 014037. [CrossRef]
37. Fernandez, A.; Baleanu, D.; Srivastava, H.M. Series representations for models of fractional calculus involving generalised Mittag-Leffler functions. *Commun. Nonlinear Sci. Numerical Simul.* **2019**, *67*, 517–527. [CrossRef]
38. Baleanu, D.; Fernandez, A. On some new properties of fractional derivatives with Mittag-Leffler kernel. *Commun. Nonlinear Sci. Numerical Simul.* **2018**, *59*, 444–462. [CrossRef]
39. Fernandez, A.; Abdeljawad, T.; Baleanu, D. Relations between fractional models with three-parameter Mittag-Leffler kernels. *Adv. Differ. Equ.* **2020**, *2020*, 186. [CrossRef]
40. Lavault, C. Integral representations and asymptotic behaviour of a Mittag- Leffler type function of two variables. *Adv. Oper. Theory* **2018**, *3*, 40–48. [CrossRef]
41. Wang, J.R.; Zhou, Y.; O'Regan, D. A note on asymptotic behaviour of Mittag–Leffler functions. *Integral Transforms Spec. Funct.* **2018**, *29*, 81–94. [CrossRef]
42. Atanacković, T.M.; Konjik, S.; Pilipović, S.; Zorica, D. Complex order fractional derivatives in viscoelasticity. *Mech. Time-Depend. Mater.* **2016**, *20*, 175–195. [CrossRef]
43. Naqvi, Q.A.; Abbas, M. Complex and higher order fractional curl operator in electromagnetics. *Opt. Commun.* **2004**, *241*, 349–355. [CrossRef]
44. Pinto, C.M.A.; Carvalho, A.R.M. Fractional complex-order model for HIV infection with drug resistance during therapy. *J. Vib. Control* **2015**, *22*, 2222–2239. [CrossRef]

© 2020 by the authors. Licensee MDPI, Basel, Switzerland. This article is an open access article distributed under the terms and conditions of the Creative Commons Attribution (CC BY) license (http://creativecommons.org/licenses/by/4.0/).

Article

On Some Formulas for the k-Analogue of Appell Functions and Generating Relations via k-Fractional Derivative

Övgü Gürel Yılmaz [1], Rabia Aktaş [2,*] and Fatma Taşdelen [2]

[1] Department of Mathematics, Recep Tayyip Erdogan University, 53100 Rize, Turkey; ovgu.gurelyilmaz@erdogan.edu.tr
[2] Department of Mathematics, Faculty of Science, Ankara University, Tandoğan, 06100 Ankara, Turkey; tasdelen@science.ankara.edu.tr
* Correspondence: raktas@science.ankara.edu.tr

Received: 21 August 2020; Accepted: 22 September 2020; Published: 24 September 2020

Abstract: Our present investigation is mainly based on the k-hypergeometric functions which are constructed by making use of the Pochhammer k-symbol in Diaz et al. 2007, which are one of the vital generalizations of hypergeometric functions. In this study, we focus on the k-analogue of F_1 Appell function introduced by Mubeen et al. 2015 and the k-generalizations of F_2 and F_3 Appell functions indicated in Kıymaz et al. 2017. We present some important transformation formulas and some reduction formulas which show close relation not only with k-Appell functions but also with k-hypergeometric functions. Employing the theory of Riemann–Liouville k-fractional derivative from Rahman et al. 2020, and using the relations which we consider in this paper, we acquire linear and bilinear generating relations for k-analogue of hypergeometric functions and Appell functions.

Keywords: k-gamma function; k-beta function; Pochhammer symbol; hypergeometric function; Appell functions; integral representation; reduction and transformation formula; fractional derivative; generating function

1. Introduction

Special functions, with their diverse sub-branches, provide a very wide field of study that appears not only in various fields of mathematics, but also in the solutions of important problems in many disciplines of science such as physics, chemistry, and biology. This subject is powerful enough to make sense of uncertain questions especially in physical problems, so it encourages many people for notable improvements on this matter. As in other sciences, remarkable problems are still discussed in many disciplines, and more general results are attempted to be obtained.

Generalized hypergeometric functions, which are some of these studies in special functions [1,2], are defined by

$$_pF_q\left[\begin{array}{c}\alpha_1,\alpha_2,...,\alpha_p\\ \beta_1,\beta_2,...,\beta_q\end{array};x\right]=\sum_{n=0}^{\infty}\frac{(\alpha_1)_n(\alpha_2)_n\cdots(\alpha_p)_n}{(\beta_1)_n(\beta_2)_n\cdots(\beta_q)_n}\frac{x^n}{n!}, \qquad (1)$$

where $\alpha_1,\alpha_2,...,\alpha_p,\beta_1,\beta_2,...,\beta_q$, $x \in \mathbb{C}$ and $\beta_1,\beta_2,...,\beta_q$ are neither zero nor negative integers.

Here, $(\lambda)_n$ is the Pochhammer symbol defined by

$$(\lambda)_n = \begin{cases} \lambda(\lambda+1)\ldots(\lambda+n-1) \;; & n \geq 1, \\ 1 \;; & n=0,\; \lambda \neq 0. \end{cases} \qquad (2)$$

For the special case that corresponds to $p = 2$ and $q = 1$ in (1), we can obtain $_2F_1$ Gauss hypergeometric function [1,2],

$$_2F_1\left[\begin{array}{c}\alpha,\beta\\ \gamma\end{array};x\right] = \sum_{n=0}^{\infty}\frac{(\alpha)_n(\beta)_n}{(\gamma)_n}\frac{x^n}{n!}, \quad |x|<1, \tag{3}$$

where $\alpha, \beta, \gamma, x \in \mathbb{C}$ and γ is neither zero nor a negative integer.

Many elementary functions can be expressed in terms of hypergeometric functions. Moreover, non-elementary functions that occur in physics and mathematics have a representation with hypergeometric series. Therefore, generalizing hypergeometric functions have many applications in mathematics and the other disciplines. For instance, in quantum field theory, hypergeometric functions appear in the calculation of the Feynmann integrals, and the analytic results can be expressed in terms of these functions [3,4]. In engineering, analytic forms of the fractional order derivatives of sinusoidal functions are represented with hypergeometric functions [5]. In biochemistry, for the analysis of a simple gene expression model, a hypergeometric probability distribution is considered [6], and they also appear in the connection and linearization problems in mathematics [7–9]. Generalization of hypergeometric function can be made by increasing the number of parameters in the hypergeometric function or by increasing the number of variables. Appell, based on the idea that the number of variables can be increased, has defined Appell hypergeometric functions obtained by multiplying two hypergeometric functions. These are the four elemanter functions defined in [1,2]

$$F_1(\alpha,\beta,\beta';\gamma;x,y) = \sum_{m,n=0}^{\infty}\frac{(\alpha)_{m+n}(\beta)_m(\beta')_n}{(\gamma)_{m+n}}\frac{x^m y^n}{m!\,n!}, \tag{4}$$

$$F_2(\alpha,\beta,\beta';\gamma,\gamma';x,y) = \sum_{m,n=0}^{\infty}\frac{(\alpha)_{m+n}(\beta)_m(\beta')_n}{(\gamma)_m(\gamma')_n}\frac{x^m y^n}{m!\,n!}, \tag{5}$$

$$F_3(\alpha,\alpha',\beta,\beta';\gamma;x,y) = \sum_{m,n=0}^{\infty}\frac{(\alpha)_m(\alpha')_n(\beta)_m(\beta')_n}{(\gamma)_{m+n}}\frac{x^m y^n}{m!\,n!}, \tag{6}$$

$$F_4(\alpha,\beta;\gamma,\gamma';x,y) = \sum_{m,n=0}^{\infty}\frac{(\alpha)_{m+n}(\beta)_{m+n}}{(\gamma)_m(\gamma')_n}\frac{x^m y^n}{m!\,n!}, \tag{7}$$

where $|x|<1, |y|<1$; $|x|+|y|<1$; $|x|<1, |y|<1$; $\sqrt{|x|}+\sqrt{|y|}<1$, respectively. Appell functions can be found in the study of autoionization of atoms [10], separability of Hamilton–Jacobi equations in classical mechanics [11], representation of Feynmann integral in quantum field theory [3,4], and expression of Nordsieck integral in atomic collisions physics [12].

Another generalization of hypergeometric functions is the hypergeometric k-function, defined by the Pochhammer k-symbol studied by Diaz et al. [13]. This paper includes the k-analogue of the Pochhammer symbol and hypergeometric function, as well as the k-generalization of gamma, beta, and zeta functions with their integral representations and some identities provided by classical ones. It should be noted that, taking $k=1$ in these generalizations, the k-extensions of the functions reduce to the classical ones.

Let $k\in\mathbb{R}^+$ and $n\in\mathbb{N}^+$. Hypergeometric k-function is defined in [13] as

$$_2F_{1,k}\left[\begin{array}{c}\alpha,\beta\\ \gamma\end{array};x\right] := {}_2F_{1,k}\left[\begin{array}{c}(\alpha,k),(\beta,k)\\ (\gamma,k)\end{array};x\right] = \sum_{n=0}^{\infty}\frac{(\alpha)_{n,k}(\beta)_{n,k}}{(\gamma)_{n,k}}\frac{x^n}{n!}, \tag{8}$$

where $\alpha, \beta, \gamma, x \in \mathbb{C}$ and γ neither zero nor a negative integer and $(\lambda)_{n,k}$ is the Pochhammer k-symbol defined in [13] as

$$(\lambda)_{n,k} = \begin{cases} \lambda(\lambda+k)(\lambda+2k)\ldots(\lambda+(n-1)k) &;\; n\geq 1,\\ 1 &;\; n=0,\,\lambda\neq 0. \end{cases} \tag{9}$$

Based on this generalization, Kokologiannaki [14] obtained different inequalities and properties for the generalizations of Gamma, Beta, and Zeta functions. Some limits with the help of asymptotic properties of k-gamma and k-beta functions were discussed by Krasniqi [15]. Mubeen et al. [16] established integral representations of the k-confluent hypergeometric function and k-hypergeometric function and, in another paper [17], proved the k-analogue of the Kummer's first formulation using these integral representations. In [18], some families of multilinear and multilateral generating functions for the k-analogue of the hypergeometric functions were obtained. Studies on this subject are not limited to these papers; for details, see [19–22].

In [23], Mubeen adapted the k-generalization to the Riemann–Liouville fractional integral by using k-gamma function. In [24], k-Riemann–Liouville fractional derivative was studied and new properties were obtained with the help of Fourier and Laplace transforms. In [25], Rahman et al. applied the newly k-fractional derivative operator to k-analogue of hypergeometric and Appell functions and obtained new relations satisfied between them. Furthermore, k-fractional derivative operator was applied to the k-Mittag–Leffler function and the Wright function.

Our present investigation is motivated by the fact that generalizations of hypergeometric functions have considerable importance due to their applications in many disciplines from different perspectives. Therefore, our study is generally based on the k-extension of hypergeometric functions. The structure of the paper is organized as follows: In Section 2, we briefly give some definitions and preliminary results which are essential in the following sections as noted in [13,23,26,27]. In Section 3, we prove some main properties such as transformation formulas, and some reduction formulas which enable us to have relations for k-hypergeometric functions and k-Appell functions. In the last part of the paper, applying the theory of Riemann–Liouville k-fractional derivative [25] and using the relations which we consider in the previous sections, we gain linear and bilinear generating relations for k-analogue of hypergeometric functions and k-Appell functions.

2. Some Definitions and Preliminary Results

For the sake of completeness, it will be better to examine the preliminary section in three subsections by the reason of the number of theorems and definitions. In these subsections, we will present some definitions, properties, and results which we need in our investigation in further sections. We begin by introducing k-gamma, k-beta, and k-analogue of hypergeometric function and we continue definitions of k-generalized F_1, F_2 and F_3 which are the classical Appell functions. We conclude this section with recalling Riemann–Liouville fractional derivative, k-generalization of this fractional derivative, and some important theorems which will be required in our studies.

Through this paper, we denote by $\mathbb{C}$, $\mathbb{R}$, $\mathbb{R}^+$, $\mathbb{N}$ and $\mathbb{N}^+$ the sets of complex numbers, real numbers, real and positive numbers, and positive integers with zero and positive integers, respectively.

2.1. k-Generalizations of Gamma, Beta, and Hypergeometric Functions

In this subsection, the definitions of k-gamma and k-beta functions are presented and some elemental relations provided by these functions are introduced by Diaz et al. [13] and Mubeen et al. [22]. Furthermore, we continue the definition of k-hypergeometric function and we present integral representation and some formulas satisfied from this generalization [16,17].

Definition 1. *For $x \in \mathbb{C}$ and $k \in \mathbb{R}^+$, the integral representation of k-gamma function Γ_k is defined by*

$$\Gamma_k(x) = \int_0^\infty t^{x-1} e^{-\frac{t^k}{k}} dt, \tag{10}$$

where $\Re(x) > 0$ [13,22].

Definition 2. For $x, y \in \mathbb{C}$ and $k \in \mathbb{R}^+$, the k-beta function B_k is defined by

$$B_k(x,y) = \frac{1}{k}\int_0^1 t^{\frac{x}{k}-1}(1-t)^{\frac{y}{k}-1}dt, \tag{11}$$

where $\Re(x) > 0$ and $\Re(y) > 0$ [13].

Proposition 1. Let $k \in \mathbb{R}^+$, $a \in \mathbb{R}$, $n \in \mathbb{N}^+$. The k-gamma function Γ_k and the k-beta function B_k satisfy the following properties [13,22],

$$\Gamma_k(x+k) = x\Gamma_k(x), \tag{12}$$

$$\Gamma_k(x) = k^{\frac{x}{k}-1}\Gamma\left(\frac{x}{k}\right), \tag{13}$$

$$B_k(x,y) = \frac{\Gamma_k(x)\Gamma_k(y)}{\Gamma_k(x+y)}, \tag{14}$$

$$B_k(x,y) = \frac{1}{k}B\left(\frac{x}{k},\frac{y}{k}\right). \tag{15}$$

Definition 3. Let $x \in \mathbb{C}$, $k \in \mathbb{R}^+$ and $n \in \mathbb{N}^+$. Then, the Pochhammer k-symbol is defined in [13,22] by

$$(x)_{n,k} = x(x+k)(x+2k)\ldots(x+(n-1)k). \tag{16}$$

In particular, we denote $(x)_{0,k} := 1$.

Proposition 2. If $\alpha \in \mathbb{C}$ and $m,n \in \mathbb{N}^+$ then for $k \in \mathbb{R}^+$, we have

$$(\alpha)_{n,k} = \frac{\Gamma_k(\alpha + nk)}{\Gamma_k(\alpha)}, \tag{17}$$

$$(\alpha)_{n,k} = k^n\left(\frac{\alpha}{k}\right)_n, \tag{18}$$

$$(\alpha)_{m+n,k} = (\alpha)_{m,k}(\alpha + mk)_{n,k}, \tag{19}$$

where $(\alpha)_n$ and $(\alpha)_{n,k}$ denote the Pochhammer symbol and Pochhammer k-symbol, respectively [13,22].

Proposition 3. For any $\alpha \in \mathbb{C}$ and $k \in \mathbb{R}^+$, the following identity holds

$$\sum_{n=0}^{\infty}(\alpha)_{n,k}\frac{x^n}{n!} = (1-kx)^{-\frac{\alpha}{k}}, \tag{20}$$

where $|x| < \frac{1}{k}$ [13,22].

Theorem 1. Assume that $x \in \mathbb{C}$, $k \in \mathbb{R}^+$ and $\Re(\gamma) > \Re(\beta) > 0$, then the integral representation of the k-hypergeometric function is defined in [16] as

$$_2F_{1,k}\left[\begin{array}{c}\alpha, \beta \\ \gamma\end{array};x\right] = \frac{\Gamma_k(\gamma)}{k\Gamma_k(\beta)\Gamma_k(\gamma-\beta)}\int_0^1 t^{\frac{\beta}{k}-1}(1-t)^{\frac{\gamma-\beta}{k}-1}(1-kxt)^{-\frac{\alpha}{k}}dt. \tag{21}$$

For the following theorem, $_2F_{1,k}\left[\begin{array}{c}(\alpha,1), (\beta,k) \\ (\gamma,k)\end{array};x\right]$ is the expression of the following form [17],

$$_2F_{1,k}^*\left[\begin{array}{c}\alpha, \beta \\ \gamma\end{array};x\right] := {_2F_{1,k}}\left[\begin{array}{c}(\alpha,1), (\beta,k) \\ (\gamma,k)\end{array};x\right] = \sum_{n=0}^{\infty}\frac{(\alpha)_n(\beta)_{n,k}}{(\gamma)_{n,k}}\frac{x^n}{n!}. \tag{22}$$

Theorem 2. In [17], assume that $x \in \mathbb{C}$, $k \in \mathbb{R}^+$ and $\operatorname{Re}(\gamma - \beta) > 0$, then

$$_2F_{1,k}\left[\begin{array}{c}(\alpha,1)\ ,\ (\beta,k)\\(\gamma,k)\end{array};x\right] := \frac{\Gamma_k(\gamma)\,\Gamma_k(\gamma-\beta-k\alpha)}{\Gamma_k(\gamma-\beta)\,\Gamma_k(\gamma-k\alpha)}. \tag{23}$$

For the special case $\alpha = -n$,

$$_2F_{1,k}\left[\begin{array}{c}(-n,1)\ ,\ (\beta,k)\\(\gamma,k)\end{array};x\right] = \frac{(\gamma-\beta)_{n,k}}{(\gamma)_{n,k}}. \tag{24}$$

2.2. k-Generalizations of the Appell Functions F_1, F_2 and F_3

In 2015, k-generalization of F_1 Appell function was introduced and contiguous function relations and integral representation of this function were shown by using the fundamental relations of the Pochhammer k-symbol [26]. In 2017, k-analogues of the F_2, F_3, and F_4 were explored by Kıymaz et al. in [27] and also in that study, they provided the relations between k-analogues of Appell functions and the classical ones. Here, we remind the definitions of k-analogue of F_1, F_2 and F_3 which are the Appell functions, and integral representations which are satisfied by them [26,27].

Definition 4. In [26], let $k \in \mathbb{R}^+$, $x, y \in \mathbb{C}$, α, β, β', $\gamma \in \mathbb{C}$ and $n \in \mathbb{N}^+$. Then, the $F_{1,k}$ function with the parameters α, β, β', γ is given by

$$F_{1,k}(\alpha,\beta,\beta';\gamma;x,y) = \sum_{m,n=0}^{\infty} \frac{(\alpha)_{m+n,k}\,(\beta)_{m,k}\,(\beta')_{n,k}}{(\gamma)_{m+n,k}}\,\frac{x^m}{m!}\,\frac{y^n}{n!}, \tag{25}$$

where $\gamma \neq 0, -1, -2, \ldots$ and $|x| < \frac{1}{k}$, $|y| < \frac{1}{k}$.

Definition 5. In [27], let $k \in \mathbb{R}^+$, $x, y \in \mathbb{C}$, α, β, β', γ, $\gamma' \in \mathbb{C}$ and $m, n \in \mathbb{N}^+$. Then, the Appell k-functions are defined by

$$F_{2,k}(\alpha,\beta,\beta';\gamma,\gamma';x,y) = \sum_{m,n=0}^{\infty} \frac{(\alpha)_{m+n,k}\,(\beta)_{m,k}\,(\beta')_{n,k}}{(\gamma)_{m,k}\,(\gamma')_{n,k}}\,\frac{x^m}{m!}\,\frac{y^n}{n!}, \tag{26}$$

$$F_{3,k}(\alpha,\alpha',\beta,\beta';\gamma;x,y) = \sum_{m,n=0}^{\infty} \frac{(\alpha)_{m,k}\,(\alpha')_{n,k}\,(\beta)_{m,k}\,(\beta')_{n,k}}{(\gamma)_{m+n,k}}\,\frac{x^m}{m!}\,\frac{y^n}{n!}, \tag{27}$$

$$F_{4,k}(\alpha,\beta;\gamma,\gamma';x,y) = \sum_{m,n=0}^{\infty} \frac{(\alpha)_{m+n,k}\,(\beta)_{m+n,k}}{(\gamma)_{m,k}\,(\gamma')_{n,k}}\,\frac{x^m}{m!}\,\frac{y^n}{n!}, \tag{28}$$

where $|x| + |y| < \frac{1}{k}$; $|x| < \frac{1}{k}, |y| < \frac{1}{k}$; $\sqrt{|x|} + \sqrt{|y|} < \frac{1}{\sqrt{k}}$, respectively, and denominators are neither zero nor negative integers.

Theorem 3. In [26], assume that $k \in \mathbb{R}^+$, $x, y \in \mathbb{C}$, $\Re(\gamma) > \Re(\alpha) > 0$, then the integral representation of the k-hypergeometric function is as follows:

$$F_{1,k}(\alpha,\beta,\beta';\gamma;x,y)$$

$$= \frac{\Gamma_k(\gamma)}{k\Gamma_k(\alpha)\Gamma_k(\gamma-\alpha)} \int_0^1 t^{\frac{\alpha}{k}-1}(1-t)^{\frac{\gamma-\alpha}{k}-1}(1-kxt)^{-\frac{\beta}{k}}(1-kyt)^{-\frac{\beta'}{k}}\,dt. \tag{29}$$

Theorem 4. *In [27], let $k \in \mathbb{R}^+$. Integral representations of $F_{2,k}$ and $F_{3,k}$ have the forms of*

$$F_{2,k}(\alpha, \beta, \beta'; \gamma, \gamma'; x, y) = \frac{1}{k^2 B_k(\beta, \gamma - \beta) B_k(\beta', \gamma' - \beta')}$$
$$\times \int_0^1 \int_0^1 \frac{t^{\frac{\beta}{k}-1} s^{\frac{\beta'}{k}-1} (1-t)^{\frac{\gamma-\beta}{k}-1} (1-s)^{\frac{\gamma'-\beta'}{k}-1}}{(1-kxt-kys)^{\frac{\alpha}{k}}} dt ds, \qquad (30)$$

$$F_{3,k}(\alpha, \alpha', \beta, \beta'; \gamma; x, y) = \frac{\Gamma_k(\gamma)}{k^2 \Gamma_k(\beta) \Gamma_k(\beta') \Gamma_k(\gamma - \beta - \beta')}$$
$$\times \iint_D \frac{t^{\frac{\beta}{k}-1} s^{\frac{\beta'}{k}-1} (1-kxt)^{-\frac{\alpha}{k}} (1-kys)^{-\frac{\alpha'}{k}}}{(1-t-s)^{1-\frac{\gamma-\beta-\beta'}{k}}} dt ds, \qquad (31)$$

where $\Re(\gamma) > \Re(\beta) > 0$, $\Re(\gamma') > \Re(\beta') > 0$ and $D = \{t \geq 0, s \geq 0, t + s \leq 1\}$.

2.3. *The Riemann–Liouville k-Fractional Derivative Operator*

Fractional calculus and its applications have been intensively investigated for a long time by many researchers in numerous disciplines and attention to this subject has grown tremendously. By making use of the concept of the fractional derivatives and integrals, various extensions of them have been introduced [28–31], and authors have gained different perspectives in many areas such as engineering, physics, economics, biology, statistics [32,33]. One of the generalizations of fractional derivatives is Riemann–Liouville k-fractional derivative operator studied in [24,25,34]. Here, we remind the definition of Riemann–Liouville fractional derivative and its k-generalization and also some theorems which will be used in the further section, are displayed.

Definition 6. *In [2], the well-known Riemann–Liouville fractional derivative of order μ is described, for a function f, as follows:*

$$\mathcal{D}_z^\mu \{f(z)\} = \frac{1}{\Gamma(-\mu)} \int_0^z f(t)(z-t)^{-\mu-1} dt, \qquad (32)$$

where $\Re(\mu) < 0$.

In particular, for the case $m - 1 < \Re(\mu) < m$, where $m = 1, 2, \ldots$ (32) is written by

$$\mathcal{D}_z^\mu \{f(z)\} = \frac{d^m}{dz^m} \mathcal{D}_z^{\mu-m} \{f(z)\} \qquad (33)$$
$$= \frac{d^m}{dz^m} \left\{ \frac{1}{\Gamma(-\mu+m)} \int_0^x f(t)(z-t)^{-\mu+m-1} dt \right\}.$$

Definition 7. *In [25], the k-analogue of Riemann–Liouville fractional derivative of order μ is defined by*

$$_k\mathcal{D}_z^\mu \{f(z)\} = \frac{1}{k\Gamma_k(-\mu)} \int_0^z f(t)(z-t)^{-\frac{\mu}{k}-1} dt, \qquad (34)$$

where $\Re(\mu) < 0$ and $k \in \mathbb{R}^+$.

In particular, for the case $m - 1 < \Re(\mu) < m$, where $m = 1, 2, \ldots$, (34) is written by

$$\begin{aligned}_k\mathcal{D}_z^\mu \{f(z)\} &= \frac{d^m}{dz^m} {}_k\mathcal{D}_z^{\mu-mk} \{f(z)\} \\ &= \frac{d^m}{dz^m} \left\{ \frac{1}{k\Gamma_k(-\mu+mk)} \int_0^z f(t)(z-t)^{-\frac{\mu}{k}+m-1} dt \right\}. \end{aligned} \quad (35)$$

Theorem 5. *In [25], let $k \in \mathbb{R}^+$, $\Re(\mu) < 0$. Then, we have*

$$_k\mathcal{D}_z^\mu \left\{ z^{\frac{\eta}{k}} \right\} = \frac{z^{\frac{\eta-\mu}{k}}}{\Gamma_k(-\mu)} B_k(\eta + k, -\mu). \quad (36)$$

Theorem 6. *In [25], let $Re(\mu) > 0$ and suppose that the function $f(z)$ is analytic at the origin with its Maclaurin expansion has the power series expansion*

$$f(z) = \sum_{n=0}^\infty a_n z^n, \quad (37)$$

where $|z| < \rho$, $\rho \in \mathbb{R}^+$. Then,

$$_k\mathcal{D}_z^\mu \{f(z)\} = \sum_{n=0}^\infty a_n \, _k\mathcal{D}_z^\mu \{z^n\}. \quad (38)$$

Theorem 7. *In [25], let $k \in \mathbb{R}^+$, $\Re(\mu) > \Re(\eta) > 0$. Then, the following result holds true:*

$$_k\mathcal{D}_z^{\eta-\mu} \left\{ z^{\frac{\eta}{k}-1}(1-kz)^{-\frac{\beta}{k}} \right\} = \frac{\Gamma_k(\eta)}{\Gamma_k(\mu)} z^{\frac{\mu}{k}-1} \, {}_2F_{1,k}\left[\begin{array}{c} \beta, \eta \\ \mu \end{array}; z \right], \quad (39)$$

where $|z| < \frac{1}{k}$.

Theorem 8. *In [25], let $k \in \mathbb{R}^+$. We have the following result:*

$$_k\mathcal{D}_z^{\eta-\mu} \left\{ z^{\frac{\eta}{k}-1}(1-kaz)^{-\frac{\alpha}{k}}(1-kbz)^{-\frac{\beta}{k}} \right\} = \frac{\Gamma_k(\eta)}{\Gamma_k(\mu)} z^{\frac{\mu}{k}-1} F_{1,k}(\eta, \alpha, \beta; \mu; az, bz), \quad (40)$$

where $\Re(\mu) > \Re(\eta) > 0$, $\Re(\alpha) > 0$, $\Re(\beta) > 0$ and $\max\{|az|, |bz|\} < \frac{1}{k}$.

3. Transformation Formulas of *k*-Generalized Appell Functions

In this section, we derive some linear transformations of *k*-generalized Appell functions and give some reduction formulas involving the $_2F_{1,k}$ hypergeometric function which provide us with an opportunity to generalize widely used identities for Appell hypergeometric functions.

Theorem 9. *For $k \in \mathbb{R}^+$, $F_{1,k}$ has the following relation:*

$$\begin{aligned}&F_{1,k}(\alpha, \beta, \beta'; \gamma; x, y) \\ &= (1-kx)^{-\frac{\beta}{k}}(1-ky)^{-\frac{\beta'}{k}} F_{1,k}\left(\gamma - \alpha, \beta, \beta'; \gamma; -\frac{x}{1-kx}, -\frac{y}{1-ky} \right), \end{aligned} \quad (41)$$

where $\Re(\gamma) > \Re(\alpha) > 0$ and $\left|\frac{x}{1-kx}\right| < \frac{1}{k}$, $\left|\frac{y}{1-ky}\right| < \frac{1}{k}$, $|x| < \frac{1}{k}$, $|y| < \frac{1}{k}$.

Proof. In [26], the integral representation of $F_{1,k}$ is given by

$$F_{1,k}(\alpha, \beta, \beta'; \gamma; x, y) = \frac{1}{kB_k(\alpha, \gamma - \alpha)} \int_0^1 t^{\frac{\alpha}{k}-1}(1-t)^{\frac{\gamma-\alpha}{k}-1}(1-kxt)^{-\frac{\beta}{k}}(1-kyt)^{-\frac{\beta'}{k}} dt.$$

If we make use of the substitution $t = 1 - t_1$ in the above integral, then we can write

$$F_{1,k}(\alpha, \beta, \beta'; \gamma; x, y) = \frac{1}{kB_k(\alpha, \gamma - \alpha)}$$

$$\times \int_0^1 t_1^{\frac{\gamma-\alpha}{k}-1}(1-t_1)^{\frac{\alpha}{k}-1}(1-kx(1-t_1))^{-\frac{\beta}{k}}(1-ky(1-t_1))^{-\frac{\beta'}{k}} dt_1$$

$$= \frac{1}{kB_k(\alpha, \gamma - \alpha)}(1-kx)^{-\frac{\beta}{k}}(1-ky)^{-\frac{\beta'}{k}}$$

$$\times \int_0^1 t_1^{\frac{\gamma-\alpha}{k}-1}(1-t_1)^{\frac{\alpha}{k}-1}\left(1+\frac{kxt_1}{1-kx}\right)^{-\frac{\beta}{k}}\left(1+\frac{kyt_1}{1-ky}\right)^{-\frac{\beta'}{k}} dt_1$$

$$= (1-kx)^{-\frac{\beta}{k}}(1-ky)^{-\frac{\beta'}{k}} F_{1,k}\left(\gamma - \alpha, \beta, \beta'; \gamma; -\frac{x}{1-kx}, -\frac{y}{1-ky}\right).$$

Thus, we get the desired result. □

Theorem 10. *For $k \in \mathbb{R}^+$, we have*

$$F_{1,k}(\alpha, \beta, \beta'; \gamma; x, y) = (1-kx)^{-\frac{\alpha}{k}} F_{1,k}\left(\alpha, \gamma - \beta - \beta', \beta'; \gamma; -\frac{x}{1-kx}, -\frac{x-y}{1-kx}\right), \quad (42)$$

and

$$F_{1,k}(\alpha, \beta, \beta'; \gamma; x, y) = (1-ky)^{-\frac{\alpha}{k}} F_{1,k}\left(\alpha, \beta, \gamma - \beta - \beta'; \gamma; -\frac{y-x}{1-ky}, -\frac{y}{1-ky}\right). \quad (43)$$

Proof. By a change of variables $t = \frac{t_1}{1-kx+kt_1x}$ in the integral representation of $F_{1,k}$, we have that

$$F_{1,k}(\alpha, \beta, \beta'; \gamma; x, y) = \frac{1}{kB_k(\alpha, \gamma - \alpha)}$$

$$\times \int_0^1 t^{\frac{\alpha}{k}-1}(1-t)^{\frac{\gamma-\alpha}{k}-1}(1-kxt)^{-\frac{\beta}{k}}(1-kyt)^{-\frac{\beta'}{k}} dt$$

$$= \frac{1}{kB_k(\alpha, \gamma - \alpha)}(1-kx)^{\frac{\gamma-\alpha-\beta}{k}}$$

$$\times \int_0^1 t_1^{\frac{\alpha}{k}-1}(1-t_1)^{\frac{\gamma-\alpha}{k}-1}(1-kx+kxt_1)^{\frac{\beta+\beta'-\gamma}{k}}(1-kx+kxt_1-kyt_1)^{-\frac{\beta'}{k}} dt_1$$

$$= \frac{1}{kB_k(\alpha, \gamma - \alpha)}(1-kx)^{-\frac{\alpha}{k}}$$

$$\times \int_0^1 t_1^{\frac{\alpha}{k}-1}(1-t_1)^{\frac{\gamma-\alpha}{k}-1}\left(1+\frac{kxt_1}{1-kx}\right)^{\frac{\beta+\beta'-\gamma}{k}}\left(1+\frac{kxt_1-kyt_1}{1-kx}\right)^{-\frac{\beta'}{k}} dt_1$$

$$= (1-kx)^{-\frac{\alpha}{k}} F_{1,k}\left(\alpha, \gamma - \beta - \beta', \beta'; \gamma; -\frac{x}{1-kx}, -\frac{x-y}{1-kx}\right).$$

In the above integral, we note that using a similar argument with $t = \frac{t_1}{1-ky-kt_1y}$, one can easily obtain

$$F_{1,k}(\alpha, \beta, \beta'; \gamma; x, y) = (1-ky)^{-\frac{\alpha}{k}} F_{1,k}\left(\alpha, \beta, \gamma - \beta - \beta'; \gamma; -\frac{y-x}{1-ky}, -\frac{y}{1-ky}\right).$$

□

Theorem 11. *Letting $k \in \mathbb{R}^+$, then $F_{1,k}$ has the following relations:*

$$\begin{aligned}
& F_{1,k}(\alpha, \beta, \beta'; \gamma; x, y) \\
& = (1-kx)^{\frac{\gamma-\alpha-\beta}{k}} (1-ky)^{-\frac{\beta'}{k}} F_{1,k}\left(\gamma - \alpha, \gamma - \beta - \beta', \beta'; \gamma; x, -\frac{y-x}{1-ky}\right),
\end{aligned} \qquad (44)$$

and

$$\begin{aligned}
& F_{1,k}(\alpha, \beta, \beta'; \gamma; x, y) \\
& = (1-kx)^{-\frac{\beta}{k}} (1-ky)^{\frac{\gamma-\alpha-\beta'}{k}} F_{1,k}\left(\gamma - \alpha, \beta, \gamma - \beta - \beta'; \gamma; -\frac{y-x}{1-kx}, y\right).
\end{aligned} \qquad (45)$$

Proof. Using $t = \frac{t_1}{1-kx+kxt_1}$ and $t_1 = 1 - t_2$ in integral representation of $F_{1,k}$, we obtain

$$F_{1,k}(\alpha, \beta, \beta'; \gamma; x, y) = \frac{1}{kB_k(\alpha, \gamma - \alpha)}$$

$$\times \int_0^1 t^{\frac{\alpha}{k}-1} (1-t)^{\frac{\gamma-\alpha}{k}-1} (1-kxt)^{-\frac{\beta}{k}} (1-kyt)^{-\frac{\beta'}{k}} dt$$

$$= \frac{1}{kB_k(\alpha, \gamma - \alpha)}$$

$$\times \int_0^1 t_2^{\frac{\gamma-\alpha}{k}-1} (1-t_2)^{\frac{\alpha}{k}-1} (1-kx)^{\frac{\gamma-\alpha-\beta}{k}} (1-kxt_2)^{\frac{\beta+\beta'-\gamma}{k}} (1-ky+kyt_2-kxt_2)^{-\frac{\beta'}{k}} dt_2$$

$$= (1-kx)^{\frac{\gamma-\alpha-\beta}{k}} (1-ky)^{-\frac{\beta'}{k}} \frac{1}{kB_k(\alpha, \gamma - \alpha)}$$

$$\times \int_0^1 t_2^{\frac{\gamma-\alpha}{k}-1} (1-t_2)^{\frac{\alpha}{k}-1} (1-kxt_2)^{\frac{\beta+\beta'-\gamma}{k}} \left(1 + \frac{kyt_2 - kxt_2}{1-ky}\right)^{-\frac{\beta'}{k}} dt_2$$

$$= (1-kx)^{\frac{\gamma-\alpha-\beta}{k}} (1-ky)^{-\frac{\beta'}{k}} F_{1,k}\left(\gamma - \alpha, \gamma - \beta - \beta', \beta'; \gamma; x, -\frac{y-x}{1-ky}\right).$$

Using the same method as above, we can reach (45) easily. □

Theorem 12. *Letting $k \in \mathbb{R}^+$, then the following relations hold:*

$$F_{2,k}(\alpha, \beta, \beta'; \gamma, \gamma'; x, y)$$
$$= (1-kx)^{-\frac{\alpha}{k}} F_{2,k}\left(\alpha, \gamma-\beta, \beta'; \gamma, \gamma'; -\frac{x}{1-kx}, \frac{y}{1-kx}\right), \quad (46)$$

$$F_{2,k}(\alpha, \beta, \beta'; \gamma, \gamma'; x, y)$$
$$= (1-ky)^{-\frac{\alpha}{k}} F_{2,k}\left(\alpha, \beta, \gamma'-\beta'; \gamma, \gamma'; \frac{x}{1-ky}, -\frac{y}{1-ky}\right), \quad (47)$$

and

$$F_{2,k}(\alpha, \beta, \beta'; \gamma, \gamma'; x, y)$$
$$= (1-kx-ky)^{-\frac{\alpha}{k}} F_{2,k}\left(\alpha, \gamma-\beta, \gamma'-\beta'; \gamma, \gamma'; -\frac{x}{1-kx-ky}, -\frac{y}{1-kx-ky}\right). \quad (48)$$

Proof. By taking for the first relation $t = 1 - t_1$, for the second $s = 1 - s_1$ and, finally, for the third $t = 1 - t_1$, $s = 1 - s_1$ together in the double integral (30), we find (46), (47) and (48), respectively. These complete the proof. □

We continue with some reduction formulas for Appell functions $F_{1,k}$ and $F_{2,k}$ in terms of the $_2F_{1,k}$ generalized hypergeometric function.

Theorem 13. *Let $k \in \mathbb{R}^+$. Then, the special cases of $F_{1,k}$ and $F_{2,k}$ are as follows:*

$$F_{1,k}(\alpha, \beta, \beta'; \gamma; x, y) = (1-kx)^{-\frac{\alpha}{k}} {}_2F_{1,k}\left[\begin{array}{c} \alpha, \beta' \\ \beta+\beta' \end{array}; -\frac{x-y}{1-kx}\right], \quad (49)$$

$$F_{1,k}(\alpha, \beta, \beta'; \gamma; x, y) = (1-ky)^{-\frac{\alpha}{k}} {}_2F_{1,k}\left[\begin{array}{c} \alpha, \beta \\ \beta+\beta' \end{array}; -\frac{y-x}{1-ky}\right], \quad (50)$$

$$F_{2,k}(\alpha, \beta, \beta'; \gamma, \gamma'; x, y) = (1-kx)^{-\frac{\alpha}{k}} {}_2F_{1,k}\left[\begin{array}{c} \alpha, \beta' \\ \gamma' \end{array}; \frac{y}{1-kx}\right], \quad (51)$$

$$F_{2,k}(\alpha, \beta, \beta'; \gamma, \gamma'; x, y) = (1-ky)^{-\frac{\alpha}{k}} {}_2F_{1,k}\left[\begin{array}{c} \alpha, \beta \\ \gamma \end{array}; \frac{x}{1-ky}\right]. \quad (52)$$

Proof. Specializing (42) and (43) for $\gamma = \beta + \beta'$ and also if we set $\gamma = \beta$ and $\gamma = \beta'$ in (46) and (47), we obtain desired results, respectively. □

In the next lemma, we will prove Euler transformation for $_2F_{1,k}$ hypergeometric function, which will be used in the next theorem.

Lemma 1. *Let $x \in \mathbb{C}$, $k \in \mathbb{R}^+$. Then, we have*

$$_2F_{1,k}\left[\begin{array}{c} \alpha, \beta \\ \gamma \end{array}; x\right] = (1-kx)^{-\frac{\beta}{k}} {}_2F_{1,k}\left[\begin{array}{c} \gamma-\alpha, \beta \\ \gamma \end{array}; -\frac{x}{1-kx}\right]. \quad (53)$$

Proof. From the definition of $_2F_{1,k}$, one gets

$$(1-kx)^{-\frac{\beta}{k}} \, _2F_{1,k}\left[\begin{array}{c}\gamma-\alpha,\,\beta\\\gamma\end{array};-\frac{x}{1-kx}\right]$$

$$=(1-kx)^{-\frac{\beta}{k}}\sum_{n=0}^{\infty}\frac{(\gamma-\alpha)_{n,k}(\beta)_{n,k}}{(\gamma)_{n,k}}\frac{(-1)^n x^n}{n!\,(1-kx)^n}$$

$$=\sum_{m,n=0}^{\infty}\frac{(\gamma-\alpha)_{n,k}(\beta)_{n,k}(\beta+nk)_{m,k}}{(\gamma)_{n,k}}\frac{(-1)^n x^{m+n}}{n!\,m!}$$

$$=\sum_{m=0}^{\infty}\sum_{n=0}^{m}\frac{(\gamma-\alpha)_{n,k}(\beta)_{m,k}}{(\gamma)_{n,k}}\frac{(-1)^n x^m}{n!\,(m-n)!}. \tag{54}$$

Using the identity $(m-n)! = \frac{(-1)^n m!}{(-m)_n}$ in (54), we thus find that

$$(1-kx)^{-\frac{\beta}{k}} \, _2F_{1,k}\left[\begin{array}{c}\gamma-\alpha,\,\beta\\\gamma\end{array};-\frac{x}{1-kx}\right]$$

$$=\sum_{m=0}^{\infty}\sum_{n=0}^{m}\frac{(\gamma-\alpha)_{n,k}(-m)_n(\beta)_{m,k}}{(\gamma)_{n,k}\,n!}\frac{x^m}{m!}$$

$$=\sum_{m=0}^{\infty} {}_2F_{1,k}\left[\begin{array}{c}(-m,1),\,(\gamma-\alpha,k)\\(\gamma,k)\end{array};1\right](\beta)_{m,k}\frac{x^m}{m!}. \tag{55}$$

Making use of (24) in (55), we get the desired result. □

Theorem 14. *Let $k\in\mathbb{R}^+$. Then, we have*

$$F_{1,k}(\alpha,\beta,\beta';\gamma;x,y) = (1-ky)^{-\frac{\beta'}{k}} F_{3,k}\left(\alpha,\gamma-\alpha,\beta,\beta';\gamma;x,-\frac{y}{1-ky}\right). \tag{56}$$

Proof. Using the definition of $F_{1,k}$ defined by (25) and making use of (53), we can write

$$F_{1,k}(\alpha,\beta,\beta';\gamma;x,y) = \sum_{m=0}^{\infty}\frac{(\alpha)_{m,k}(\beta)_{m,k}}{(\gamma)_{m,k}} \, _2F_{1,k}\left[\begin{array}{c}\alpha+mk,\,\beta'\\\gamma+mk\end{array};y\right]\frac{x^m}{m!}$$

$$=\sum_{m=0}^{\infty}\frac{(\alpha)_{m,k}(\beta)_{m,k}}{(\gamma)_{m,k}}(1-ky)^{-\frac{\beta'}{k}} \, _2F_{1,k}\left[\begin{array}{c}\beta',\,\gamma-\alpha\\\gamma+mk\end{array};-\frac{y}{1-ky}\right]\frac{x^m}{m!}$$

$$=(1-ky)^{-\frac{\beta'}{k}}\sum_{m,n=0}^{\infty}\frac{(\alpha)_{m,k}(\beta)_{m,k}(\beta')_{n,k}(\gamma-\alpha)_{n,k}}{(\gamma)_{m,k}(\gamma+mk)_{n,k}}\frac{x^m}{m!}\frac{\left(-\frac{y}{1-ky}\right)^n}{n!}$$

$$=(1-ky)^{-\frac{\beta'}{k}} F_{3,k}\left(\alpha,\gamma-\alpha,\beta,\beta';\gamma;x,-\frac{y}{1-ky}\right).$$

Thus, we finish the proof. □

4. Generating Relations Involving the Generalized Appell Functions

In this section, employing the theory of Riemann–Liouville k-fractional derivative [25] and making use of the relations which we have considered in the previous sections, we establish linear and bilinear generating relations for k-analogue of hypergeometric functions and k-Appell functions.

Theorem 15. *We have the generating relation*

$$\sum_{n=0}^{\infty} \frac{(\lambda)_{n,k}}{n!} {}_2F_{1,k}\left[\begin{array}{c}\lambda+nk, \ \alpha \\ \beta\end{array};x\right]t^n = (1-kt)^{-\frac{\lambda}{k}} {}_2F_{1,k}\left[\begin{array}{c}\lambda, \ \alpha \\ \beta\end{array};\frac{x}{1-kt}\right], \quad (57)$$

where $|x| < \frac{1}{k}\min\{1, 1-kt\}$.

Proof. To prove the result, consider the elementary identities given by

$$(1-kx-kt)^{-\frac{\lambda}{k}} = (1-kt)^{-\frac{\lambda}{k}}\left(1 - \frac{kx}{1-kt}\right)^{-\frac{\lambda}{k}}, \quad (58)$$

$$(1-kx-kt)^{-\frac{\lambda}{k}} = (1-kx)^{-\frac{\lambda}{k}}\left(1 - \frac{kt}{1-kx}\right)^{-\frac{\lambda}{k}}.$$

From the series expansion using the definition of the Pochhammer k-symbol [13]

$$\sum_{n=0}^{\infty} (\alpha)_{n,k}\frac{z^n}{n!} = (1-kz)^{-\frac{\alpha}{k}},$$

We can write

$$(1-kx-kt)^{-\frac{\lambda}{k}} = (1-kx)^{-\frac{\lambda}{k}}\sum_{n=0}^{\infty}\frac{(\lambda)_{n,k}}{n!}\left(\frac{t}{1-kx}\right)^n$$

$$= (1-kx)^{-\frac{\lambda}{k}}\sum_{n=0}^{\infty}\frac{(\lambda)_{n,k}}{n!}(1-kx)^{-n}t^n$$

$$= \sum_{n=0}^{\infty}\frac{(\lambda)_{n,k}}{n!}(1-kx)^{-\frac{\lambda}{k}-n}t^n. \quad (59)$$

From (58) and (59), we have the equality

$$\sum_{n=0}^{\infty}\frac{(\lambda)_{n,k}}{n!}(1-kx)^{-\frac{\lambda}{k}-n}t^n = (1-kt)^{-\frac{\lambda}{k}}\left(1 - \frac{kx}{1-kt}\right)^{-\frac{\lambda}{k}}, \quad (60)$$

where $|t| < |1-kx|$. Multiplying both sides of (60) by $x^{\frac{\alpha}{k}-1}$ and then applying ${}_kD_x^{\alpha-\beta}$ to both sides of (60), we can reach

$${}_kD_x^{\alpha-\beta}\left\{\sum_{n=0}^{\infty}\frac{(\lambda)_{n,k}}{n!}x^{\frac{\alpha}{k}-1}(1-kx)^{-\frac{\lambda}{k}-n}t^n\right\} = {}_kD_x^{\alpha-\beta}\left\{(1-kt)^{-\frac{\lambda}{k}}x^{\frac{\alpha}{k}-1}\left(1-\frac{kx}{1-kt}\right)^{-\frac{\lambda}{k}}\right\}.$$

Since $\Re(\alpha) > 0$ where $|t| < |1-kx|$, it is possible to change the order of the summation and differentiation, we get

$$\sum_{n=0}^{\infty}\frac{(\lambda)_{n,k}}{n!} {}_kD_x^{\alpha-\beta}\left\{x^{\frac{\alpha}{k}-1}(1-kx)^{-\frac{\lambda}{k}-n}\right\}t^n \quad (61)$$

$$= (1-kt)^{-\frac{\lambda}{k}} {}_kD_x^{\alpha-\beta}\left\{x^{\frac{\alpha}{k}-1}\left(1-\frac{kx}{1-kt}\right)^{-\frac{\lambda}{k}}\right\}.$$

Finally, using relation (39) in (61), it follows that

$$\sum_{n=0}^{\infty} \frac{(\lambda)_{n,k}}{n!} {}_2F_{1,k}\left[\begin{array}{cc}\lambda+nk, & \alpha \\ & \beta\end{array};x\right]t^n = (1-kt)^{-\frac{\lambda}{k}} {}_2F_{1,k}\left[\begin{array}{cc}\lambda, & \alpha \\ & \beta\end{array};\frac{x}{1-kt}\right],$$

where $|x| < \frac{1}{k} \min\{1, 1-kt\}$. Hence, we get the desired result. □

Theorem 16. *We have the generating relation*

$$\sum_{n=0}^{\infty} \frac{(\lambda)_{n,k}}{n!} {}_2F_{1,k}\left[\begin{array}{cc}\rho-nk, & \alpha \\ & \beta\end{array};x\right]t^n$$
$$= (1-kt)^{-\frac{\lambda}{k}} F_{1,k}\left[\alpha, \rho, \lambda; \beta; x, -\frac{kxt}{1-kt}\right], \tag{62}$$

where $|x| < \frac{1}{k}$, $\left|\frac{kxt}{1-kt}\right| < \frac{1}{k}$.

Proof. Consider the identity

$$(1-k(1-kx)t)^{-\frac{\lambda}{k}} = (1-kt)^{-\frac{\lambda}{k}}\left(1+\frac{k^2xt}{1-kt}\right)^{-\frac{\lambda}{k}}. \tag{63}$$

Under the assumption $|kt| < |1-kx|^{-1}$, we can rewrite (63)

$$\sum_{n=0}^{\infty} \frac{(\lambda)_{n,k}}{n!}(1-kx)^n t^n = (1-kt)^{-\frac{\lambda}{k}}\left(1+\frac{k^2xt}{1-kt}\right)^{-\frac{\lambda}{k}}. \tag{64}$$

Multiplying $x^{\frac{\alpha}{k}-1}(1-kx)^{-\frac{\rho}{k}}$ and taking the ${}_kD_x^{\alpha-\beta}$ on both sides of (64), we obtain

$${}_kD_x^{\alpha-\beta}\left\{\sum_{n=0}^{\infty} \frac{(\lambda)_{n,k}}{n!} x^{\frac{\alpha}{k}-1}(1-kx)^{n-\frac{\rho}{k}} t^n\right\}$$
$$= {}_kD_x^{\alpha-\beta}\left\{x^{\frac{\alpha}{k}-1}(1-kt)^{-\frac{\lambda}{k}}(1-kx)^{-\frac{\rho}{k}}\left(1+k\frac{xt}{1-kt}\right)^{-\frac{\lambda}{k}}\right\}.$$

For $\Re(\alpha) > 0$, interchanging the order of the summation and the operator ${}_kD_x^{\alpha-\beta}$, we have

$$\sum_{n=0}^{\infty} \frac{(\lambda)_{n,k}}{n!} {}_kD_x^{\alpha-\beta}\left\{x^{\frac{\alpha}{k}-1}(1-kx)^{n-\frac{\rho}{k}}\right\} t^n$$
$$= (1-kt)^{-\frac{\lambda}{k}} {}_kD_x^{\alpha-\beta}\left\{x^{\frac{\alpha}{k}-1}(1-kx)^{-\frac{\rho}{k}}\left(1+k\frac{xt}{1-kt}\right)^{-\frac{\lambda}{k}}\right\}.$$

Assuming $|x| < \frac{1}{k}$ and $\left|\frac{kxt}{1-kt}\right| < \frac{1}{k}$ and using (39) and (40),

$$\sum_{n=0}^{\infty} \frac{(\lambda)_{n,k}}{n!} {}_2F_{1,k}\left[\begin{array}{cc}\rho-nk, & \alpha \\ & \beta\end{array};x\right]t^n = (1-kt)^{-\frac{\lambda}{k}} F_{1,k}\left[\alpha,\rho,\lambda;\beta;x,-\frac{kxt}{1-kt}\right],$$

the theorem is immediate. □

Theorem 17. We have the generating relations

$$\sum_{n=0}^{\infty} \frac{(\beta - \rho)_{n,k}}{n!} {}_2F_{1,k} \left[\begin{array}{c} \rho - nk, \ \alpha \\ \beta \end{array} ; x \right] t^n$$
$$= (1 - kt)^{\frac{\alpha + \rho - \beta}{k}} \left(1 - kt + k^2 xt\right)^{-\frac{\alpha}{k}} {}_2F_{1,k} \left[\begin{array}{c} \alpha, \rho \\ \beta \end{array} ; \frac{x}{1 - kt + k^2 xt} \right], \tag{65}$$

and

$$\sum_{n=0}^{\infty} \frac{(\beta)_{n,k} (\gamma)_{n,k}}{(\delta)_{n,k} \, n!} {}_2F_{1,k} \left[\begin{array}{c} -nk, \ \alpha \\ \beta \end{array} ; x \right] t^n$$
$$= F_{1,k}(\gamma, \beta - \alpha, \alpha; \delta; t, (1 - kx)t). \tag{66}$$

Proof. We use the result of the previous theorem. Setting $\lambda = \beta - \rho$ in (62), we find that

$$\sum_{n=0}^{\infty} \frac{(\beta - \rho)_{n,k}}{n!} {}_2F_{1,k} \left[\begin{array}{c} \rho - nk, \ \alpha \\ \beta \end{array} ; x \right] t^n = (1 - kt)^{\frac{\rho - \beta}{k}} F_{1,k} \left[\alpha, \rho, \beta - \rho; \beta; x, -\frac{kxt}{1 - kt} \right].$$

If we use the reduction formula for $F_{1,k}$ given by (50), we can easily obtain the desired result as follows:

$$\sum_{n=0}^{\infty} \frac{(\beta - \rho)_{n,k}}{n!} {}_2F_{1,k} \left[\begin{array}{c} \rho - nk, \ \alpha \\ \beta \end{array} ; x \right] t^n$$
$$= (1 - kt)^{\frac{\alpha + \rho - \beta}{k}} \left(1 - kt + k^2 xt\right)^{-\frac{\alpha}{k}} {}_2F_{1,k} \left[\begin{array}{c} \alpha, \rho \\ \beta \end{array} ; \frac{x}{1 - kt + k^2 xt} \right]. \tag{67}$$

For $\rho = 0$, (67) gives

$$\sum_{n=0}^{\infty} \frac{(\beta)_{n,k}}{n!} {}_2F_{1,k} \left[\begin{array}{c} -nk, \ \alpha \\ \beta \end{array} ; x \right] t^n = (1 - kt)^{\frac{\alpha - \beta}{k}} \left(1 - kt + k^2 xt\right)^{-\frac{\alpha}{k}}. \tag{68}$$

Multiplying both sides of (68) with $t^{\frac{\gamma}{k}-1}$ and operation of the ${}_kD_t^{\gamma-\delta}$ on (68), one can easily obtain

$$\sum_{n=0}^{\infty} \frac{(\beta)_{n,k}}{n!} {}_2F_{1,k} \left[\begin{array}{c} -nk, \ \alpha \\ \beta \end{array} ; x \right] {}_kD_t^{\gamma-\delta} \left\{ t^{n+\frac{\gamma}{k}-1} \right\}$$
$$= {}_kD_t^{\gamma-\delta} \left\{ t^{\frac{\gamma}{k}-1}(1 - kt)^{\frac{\alpha-\beta}{k}} \left(1 - kt + k^2 xt\right)^{-\frac{\alpha}{k}} \right\}. \tag{69}$$

In view of (36) and (40) on the right and left sides of (69), respectively, we can reach

$$\sum_{n=0}^{\infty} \frac{(\beta)_{n,k} (\gamma)_{n,k}}{(\delta)_{n,k} \, n!} {}_2F_{1,k} \left[\begin{array}{c} -nk, \ \alpha \\ \beta \end{array} ; x \right] t^n = F_{1,k}(\gamma, \beta - \alpha, \alpha; \delta; t, (1 - kx)t).$$

□

Theorem 18. We have the generating relation

$$\sum_{n=0}^{\infty} \frac{(\lambda)_{n,k}}{n!} {}_2F_{1,k} \left[\begin{array}{c} \lambda + nk, \ \alpha \\ \beta \end{array} ; x \right] {}_2F_{1,k} \left[\begin{array}{c} -nk, \ \gamma \\ \delta \end{array} ; y \right] t^n$$
$$= (1 - kt)^{-\frac{\lambda}{k}} F_{2,k} \left(\lambda, \alpha, \gamma; \beta, \delta; \frac{x}{1 - kt}, -\frac{kyt}{1 - kt} \right). \tag{70}$$

Proof. Putting $(1 - ky)\,t$ instead of t in (57), we can obtain

$$\sum_{n=0}^{\infty} \frac{(\lambda)_{n,k}}{n!} \, {}_2F_{1,k}\left[\begin{matrix} \lambda + nk, & \alpha \\ & \beta \end{matrix} ; x \right] (1-ky)^n \, t^n$$

$$= (1 - k(1-ky)\,t)^{-\frac{\lambda}{k}} \, {}_2F_{1,k}\left[\begin{matrix} \lambda, & \alpha \\ & \beta \end{matrix} ; \frac{x}{1 - k(1-ky)\,t} \right]. \tag{71}$$

Multiplying with $y^{\frac{\gamma}{k}-1}$, employing ${}_kD_y^{\gamma-\delta}$ both sides of (71) and the under the assumption $\Re(\gamma) > 0$ interchanging differentiation and summation, we can write

$$\sum_{n=0}^{\infty} \frac{(\lambda)_{n,k}}{n!} \, {}_2F_{1,k}\left[\begin{matrix} \lambda + nk, & \alpha \\ & \beta \end{matrix} ; x \right] {}_kD_y^{\gamma-\delta}\left\{ y^{\frac{\gamma}{k}-1}(1-ky)^n \right\} t^n \tag{72}$$

$$= {}_kD_y^{\gamma-\delta}\left\{ y^{\frac{\gamma}{k}-1}(1 - k(1-ky)\,t)^{-\frac{\lambda}{k}} \, {}_2F_{1,k}\left[\begin{matrix} \lambda, & \alpha \\ & \beta \end{matrix} ; \frac{x}{1-k(1-ky)t} \right]\right\}.$$

Make use of the formula (39), we can easily simplify left side of the (72) as follows:

$$\sum_{n=0}^{\infty} \frac{(\lambda)_{n,k}}{n!} \, {}_2F_{1,k}\left[\begin{matrix} \lambda + nk, & \alpha \\ & \beta \end{matrix} ; x \right] {}_kD_y^{\gamma-\delta}\left\{ y^{\frac{\gamma}{k}-1}(1-ky)^n \right\} t^n \tag{73}$$

$$= \frac{\Gamma_k(\gamma)}{\Gamma_k(\delta)} y^{\frac{\delta}{k}-1} \sum_{n=0}^{\infty} \frac{(\lambda)_{n,k}}{n!} \, {}_2F_{1,k}\left[\begin{matrix} \lambda + nk, & \alpha \\ & \beta \end{matrix} ; x \right] {}_2F_{1,k}\left[\begin{matrix} -nk, & \gamma \\ & \delta \end{matrix} ; y \right] t^n.$$

For the right side of the (72), using the definition of ${}_2F_{1,k}$ and the formula (36), one can obtain

$${}_kD_y^{\gamma-\delta}\left\{ y^{\frac{\gamma}{k}-1}(1-k(1-ky)\,t)^{-\frac{\lambda}{k}} \, {}_2F_{1,k}\left[\begin{matrix} \lambda, & \alpha \\ & \beta \end{matrix} ; \frac{x}{1-k(1-ky)\,t} \right]\right\}$$

$$= \frac{\Gamma_k(\gamma)}{\Gamma_k(\delta)} (1-kt)^{-\frac{\lambda}{k}} y^{\frac{\delta}{k}-1} F_{2,k}\left(\lambda, \alpha, \gamma; \beta, \delta; \frac{x}{1-kt}, -\frac{kyt}{1-kt} \right), \tag{74}$$

where $|x| < \frac{1}{k}$, $|y| < \frac{1}{k}$, $\left|\frac{x}{1-kt}\right| + \left|\frac{kyt}{1-kt}\right| < \frac{1}{k}$, $\left|\frac{1-ky}{1-x}t\right| < \frac{1}{k}$. Combining the relations (73) and (74), we get desired result. □

As a special case of (70), we give the following theorem.

Theorem 19. *We have the generating relation*

$$\sum_{n=0}^{\infty} \frac{(\beta-\rho)_{n,k}}{n!} \, {}_2F_{1,k}\left[\begin{matrix} \rho - nk, & \alpha \\ & \beta \end{matrix} ; x \right] {}_2F_{1,k}\left[\begin{matrix} -nk, & \gamma \\ & \delta \end{matrix} ; y \right] t^n \tag{75}$$

$$= (1-kx)^{-\frac{\alpha}{k}}(1-kt)^{\frac{\rho-\beta}{k}} F_{2,k}\left(\beta-\rho, \alpha, \gamma; \beta, \delta; -\frac{x}{(1-kx)(1-kt)}, -\frac{kyt}{1-kt} \right).$$

Proof. For $\lambda = \beta - \rho$ in (70), we get

$$\sum_{n=0}^{\infty} \frac{(\beta-\rho)_{n,k}}{n!} \, {}_2F_{1,k}\left[\begin{matrix} \beta - \rho + nk, & \alpha \\ & \beta \end{matrix} ; x \right] {}_2F_{1,k}\left[\begin{matrix} -nk, & \gamma \\ & \delta \end{matrix} ; y \right] t^n$$

$$= (1-kt)^{\frac{\rho-\beta}{k}} F_{2,k}\left(\beta-\rho, \alpha, \gamma; \beta, \delta; \frac{x}{1-kt}, -\frac{kyt}{1-kt} \right).$$

Using Euler transformation given by (53) for $_2F_{1,k}$

$$\sum_{n=0}^{\infty} \frac{(\beta-\rho)_{n,k}}{n!} (1-kx)^{-\frac{\alpha}{k}} {}_2F_{1,k}\left[\begin{array}{cc}\rho-nk, & \alpha \\ \beta & \end{array}; -\frac{x}{1-kx}\right] {}_2F_{1,k}\left[\begin{array}{cc}-nk, & \gamma \\ \delta & \end{array}; y\right] t^n$$

$$= (1-kt)^{\frac{\rho-\beta}{k}} F_{2,k}\left(\beta-\rho,\alpha,\gamma;\beta,\delta;\frac{x}{1-kt},-\frac{kyt}{1-kt}\right),$$

and putting $-\frac{x}{1-kx}$ instead of x, we reach the desired result. □

Theorem 20. *We have the generating relation*

$$\sum_{n=0}^{\infty} \frac{(\lambda)_{n,k}}{n!} {}_2F_{1,k}\left[\begin{array}{cc}\lambda+nk, & \alpha \\ \beta & \end{array}; x\right] {}_2F_{1,k}\left[\begin{array}{cc}\lambda+nk, & \gamma \\ \delta & \end{array}; y\right] t^n$$

$$= (1-kt)^{-\frac{\lambda}{k}} \sum_{n=0}^{\infty} \frac{(\lambda)_{n,k}(\alpha)_{n,k}}{(\beta)_{n,k} n!} \left(-\frac{kxy}{1-kt}\right)^n \quad (76)$$

$$\times F_{2,k}\left(\lambda+nk,\alpha+nk,\gamma+nk;\beta+nk,\delta+nk;\frac{x}{1-kt},-\frac{ky}{1-kt}\right).$$

Proof. Replacing t by $\frac{t}{1-ky}$ and after some simplification in (57), we find that

$$\sum_{n=0}^{\infty} \frac{(\lambda)_{n,k}}{n!} {}_2F_{1,k}\left[\begin{array}{cc}\lambda+nk, & \alpha \\ \beta & \end{array}; x\right] \frac{t^n}{(1-ky)^{n+\frac{\lambda}{k}}}$$

$$= (1-kt)^{-\frac{\lambda}{k}} \sum_{n=0}^{\infty} \frac{(\lambda)_{n,k}(\alpha)_{n,k}}{(\beta)_{n,k} n!} \left(\frac{x(1-ky)}{1-kt}\right)^n \left(1-\frac{ky}{1-kt}\right)^{-n-\frac{\lambda}{k}}.$$

Using the binomial expansion $(x+y)^n = \sum_{k=0}^{n} \binom{n}{k} x^k y^{n-k}$,

$$\sum_{n=0}^{\infty} \frac{(\lambda)_{n,k}}{n!} {}_2F_{1,k}\left[\begin{array}{cc}\lambda+nk, & \alpha \\ \beta & \end{array}; x\right] \frac{t^n}{(1-ky)^{n+\frac{\lambda}{k}}}$$

$$= (1-kt)^{-\frac{\lambda}{k}}$$

$$\times \sum_{n=0}^{\infty} \sum_{k_1=0}^{n} \frac{(\lambda)_{n,k}(\alpha)_{n,k}}{(\beta)_{n,k} n!} \binom{n}{k_1} (-1)^{n-k_1} \left(\frac{x}{1-kt}\right)^{k_1} \left(\frac{xky}{1-kt}\right)^{n-k_1} \left(1-\frac{ky}{1-kt}\right)^{-n-\frac{\lambda}{k}}$$

$$= (1-kt)^{-\frac{\lambda}{k}}$$

$$\times \sum_{n,k_1=0}^{\infty} \frac{(\lambda)_{n+k_1,k}(\alpha)_{n+k_1,k}}{(\beta)_{n+k_1,k} (n+k_1)!} \binom{n+k_1}{k_1} (-1)^n \left(\frac{x}{1-kt}\right)^{k_1} \left(\frac{xky}{1-kt}\right)^n \left(1-\frac{ky}{1-kt}\right)^{-n-k_1-\frac{\lambda}{k}}$$

$$= (1-kt)^{-\frac{\lambda}{k}}$$

$$\times \sum_{n=0}^{\infty} \frac{(\lambda)_{n,k}(\alpha)_{n,k}}{(\beta)_{n,k} n!} \left(-\frac{xky}{1-kt}\right)^n \left(1-\frac{ky}{1-kt}\right)^{-n-\frac{\lambda}{k}} {}_2F_{1,k}\left[\begin{array}{cc}\lambda+nk, & \alpha+nk \\ \beta+nk & \end{array}; \frac{\frac{x}{1-kt}}{1-\frac{ky}{1-kt}}\right] \quad (77)$$

Multiplying $y^{\frac{\gamma}{k}-1}$, operating ${}_kD_y^{\gamma-\delta}$ and applying (36), (39), and (40) together both sides of the (77) (in a similar way of proof of the (70)) for $|x| < \frac{1}{k}$, $|y| < \frac{1}{k}$, $\left|\frac{x}{1-kt}\right| + \left|\frac{ky}{1-kt}\right| < \frac{1}{k}$, we complete the proof. □

Theorem 21. *We have the generating relation*

$$\sum_{n=0}^{\infty} \frac{(\lambda)_{n,k}}{n!} \, {}_2F_{1,k}\left[\begin{array}{cc} \lambda+nk, & \alpha \\ & \beta \end{array}; x\right] {}_2F_{1,k}\left[\begin{array}{cc} \lambda+nk, & \gamma \\ & \delta \end{array}; y\right] t^n$$

$$= (1-kt)^{-\frac{\lambda}{k}} \sum_{n=0}^{\infty} \frac{(\lambda)_{n,k}(\alpha)_{n,k}(\gamma)_{n,k}}{(\beta)_{n,k}(\delta)_{n,k} n!} \left(\frac{k^3 xyt}{(1-kt)^2}\right)^n \quad (78)$$

$$\times {}_2F_{1,k}\left[\begin{array}{cc} \lambda+nk, & \alpha+nk \\ & \beta+nk \end{array}; \frac{x}{1-kt}\right] {}_2F_{1,k}\left[\begin{array}{cc} \lambda+nk, & \gamma+nk \\ & \delta+nk \end{array}; \frac{y}{1-kt}\right].$$

For the special case $\beta = \delta = \lambda$, *we have*

$$\sum_{n=0}^{\infty} \frac{(\lambda)_{n,k}}{n!} \, {}_2F_{1,k}\left[\begin{array}{cc} \lambda+nk, & \alpha \\ & \lambda \end{array}; x\right] {}_2F_{1,k}\left[\begin{array}{cc} \lambda+nk, & \gamma \\ & \lambda \end{array}; y\right] t^n$$

$$= (1-kt)^{\frac{\gamma+\alpha-\lambda}{k}} (1-kt-kx)^{-\frac{\alpha}{k}} (1-kt-ky)^{-\frac{\gamma}{k}}$$

$$\times {}_2F_{1,k}\left[\begin{array}{cc} \alpha, & \gamma \\ & \lambda \end{array}; \frac{k^3 xyt}{(1-kt-kx)(1-kt-ky)}\right]. \quad (79)$$

Proof. From the elementary identity, we find that

$$((1-kx)(1-ky) - kt)^{-\frac{\lambda}{k}} = (1-kt)^{-\frac{\lambda}{k}} \left(\left(1-\frac{kx}{1-kt}\right)\left(1-\frac{ky}{1-kt}\right) - \frac{k^3 xyt}{(1-kt)^2}\right)^{-\frac{\lambda}{k}}, \quad (80)$$

for $\left|\frac{kt}{(1-kx)(1-ky)}\right| < \frac{1}{k}$ and $\left|\frac{k^3 xyt}{(1-kt-kx)(1-kt-ky)}\right| < \frac{1}{k}$. Applying (20) to (80), multiplying $x^{\frac{\alpha}{k}-1} y^{\frac{\gamma}{k}-1}$ and taking ${}_kD_x^{\alpha-\beta} \, {}_kD_y^{\gamma-\delta}$ together both sides of (80), we have

$${}_kD_x^{\alpha-\beta} \, {}_kD_y^{\gamma-\delta} \left\{ \sum_{n=0}^{\infty} \frac{(\lambda)_{n,k}}{n!} x^{\frac{\alpha}{k}-1} (1-kx)^{-\frac{\lambda}{k}-n} y^{\frac{\gamma}{k}-1} (1-ky)^{-\frac{\lambda}{k}-n} t^n \right\}$$

$$= (1-kt)^{-\frac{\lambda}{k}}$$

$$\times {}_kD_x^{\alpha-\beta} \, {}_kD_y^{\gamma-\delta} \left\{ \sum_{n=0}^{\infty} \frac{(\lambda)_{n,k} (k^3 t)^n}{n!(1-kt)^{2n}} x^{\frac{\alpha}{k}+n-1} \left(1-\frac{kx}{1-kt}\right)^{-\frac{\lambda}{k}-n} y^{\frac{\gamma}{k}+n-1} \left(1-\frac{ky}{1-kt}\right)^{-\frac{\lambda}{k}-n} \right\}.$$

Under the conditions $\Re(\alpha) > 0$, $\Re(\gamma) > 0$, $|x| < \frac{1}{k}$, $|y| < \frac{1}{k}$, $\left|\frac{x}{1-kt}\right| < \frac{1}{k}$ and $\left|\frac{y}{1-kt}\right| < \frac{1}{k}$, directly from the properties (36), (39), and (40), we can obtain

$$\sum_{n=0}^{\infty} \frac{(\lambda)_{n,k}}{n!} \, {}_2F_{1,k}\left[\begin{array}{cc} \lambda+nk, & \alpha \\ & \beta \end{array}; x\right] {}_2F_{1,k}\left[\begin{array}{cc} \lambda+nk, & \gamma \\ & \delta \end{array}; y\right] t^n$$

$$= (1-kt)^{-\frac{\lambda}{k}} \sum_{n=0}^{\infty} \frac{(\lambda)_{n,k}(\alpha)_{n,k}(\gamma)_{n,k}}{(\beta)_{n,k}(\delta)_{n,k} n!} \left(\frac{k^3 xyt}{(1-kt)^2}\right)^n$$

$$\times {}_2F_{1,k}\left[\begin{array}{cc} \lambda+nk, & \alpha+nk \\ & \beta+nk \end{array}; \frac{x}{1-kt}\right] {}_2F_{1,k}\left[\begin{array}{cc} \lambda+nk, & \gamma+nk \\ & \delta+nk \end{array}; \frac{y}{1-kt}\right].$$

For the special case $\beta = \delta = \lambda$ in (78), we have

$$\sum_{n=0}^{\infty} \frac{(\lambda)_{n,k}}{n!} {}_2F_{1,k}\left[\begin{array}{c} \lambda + nk, \ \alpha \\ \lambda \end{array}; x\right] {}_2F_{1,k}\left[\begin{array}{c} \lambda + nk, \ \gamma \\ \lambda \end{array}; y\right] t^n$$

$$= (1-kt)^{-\frac{\lambda}{k}}$$

$$\times \sum_{n=0}^{\infty} \frac{(\alpha)_{n,k}(\gamma)_{n,k}}{(\lambda)_{n,k} n!} \left(\frac{k^3 xyt}{(1-kt)^2}\right)^n \left(1-\frac{kx}{1-kt}\right)^{-\frac{\alpha+nk}{k}} \left(1-\frac{ky}{1-kt}\right)^{-\frac{\gamma+nk}{k}}$$

$$= (1-kt)^{\frac{\gamma+\alpha-\lambda}{k}} (1-kt-kx)^{-\frac{\alpha}{k}} (1-kt-ky)^{-\frac{\gamma}{k}}$$

$$\times {}_2F_{1,k}\left[\begin{array}{c} \alpha, \ \gamma \\ \lambda \end{array}; \frac{k^3 xyt}{(1-kt-kx)(1-kt-ky)}\right].$$

□

5. Conclusions

Hypergeometric functions play an important role in many disciplines from different perspectives. Therefore, generalizations of hypergeometric functions have considerable popularity in many fields of science. This work is generally based on the k-extension of hypergeometric functions. By making use of the concept of the [26,27], we focus on the generalization of the Appell functions and present some transformation and reduction formulas. Using the theory of Riemann–Liouville k-fractional derivative and combining this theory with the Appell functions, we derive some linear and bilinear generating functions.

Author Contributions: Investigation Ö.G.Y., R.A., and F.T.; writing—original draft Ö.G.Y. and R.A.; writing—review and editing F.T. All authors have read and agreed to the published version of the manuscript.

Funding: This research received no external funding.

Conflicts of Interest: The authors declare no conflict of interest.

References

1. Slater, L.J. *Generalized Hypergeometric Functions*; Cambridge University Press: Cambridge, UK, 1966.
2. Srivastava, H.; Manocha, H. *Treatise on Generating Functions*; John Wiley & Sons, Inc.: New York, NY, USA, 1984.
3. Berends, F.A.; Davydychev, A.I.; Smirnov, V.A. Small-threshold behaviour of two-loop self-energy diagrams: Two-particle thresholds. *Nucl. Phys. B* **1996**, *478*, 59–89. [CrossRef]
4. Shpot, M.A. A massive Feynman integral and some reduction relations for Appell functions. *J. Math. Phys.* **2007**, *48*, 12. [CrossRef]
5. Włodarczyk, M.; Zawadzki, A. The application of hypergeometric functions to computing fractional order derivatives of sinusoidal functions. *Bull. Pol. Acad. Sci. Tech. Sci.* **2016**, *64*, 17. [CrossRef]
6. Chipindirwi, S. Analysis of A Simple Gene Expression Model. M.S. Thesis, University of Zimbabwe, Harare, Zimbabwe, 2012.
7. Abd-Elhameed, W.M. Linearization coefficients of some particular Jacobi polynomials via hypergeometric functions. *Adv. Differ. Equ.* **2016**, *2016*, 1–13. [CrossRef]
8. Fröhlich, J. Parameter derivatives of the Jacoby polynomials and the gaussian hypergeometric function. *Integral Transform. Spec. Funct.* **1994**, *2*, 253–266. [CrossRef]
9. Lewanowicz, S. *The Hypergeometric Functions Approach to the Connection Problem for the Classical Orthogonal Polynomials*; University of Wroclaw: Wroclaw, Poland, 2003.
10. Ancarani, L.U.; Del Punta, J.A.; Gasaneo, G. Derivatives of Horn hypergeometric functions with respect to their parameters. *J. Math. Phys.* **2017**, *58*, 127–165. [CrossRef]
11. Dragovic, V. The Appell hypergeometric functions and classical separable mechanical systems. *J. Phys. Math. Gen.* **2002**, *35*, 84–97. [CrossRef]

12. Colavecchia, F.D.; Gasaneo, G.; Garibotti, C.R. Hypergeometric integrals arising in atomic collisions physics. *J. Math. Phys.* **1997**, *38*, 567–583. [CrossRef]
13. Diaz, R.; Pariguan, E. On hypergeometric functions and pochhammer *k*-symbol. *Divulg. Mat.* **2007**, *15*, 179–192.
14. Kokologiannaki, C.G. Properties and inequalities of generalized *k*-gamma, beta and zeta functions. *Int. Contemp. Math. Sci.* **2010**, *5*, 653–660.
15. Krasniqi, V. A limit for the *k*-gamma and *k*-beta function. *Int. Math. Forum.* **2010**, *5*, 1613–1617.
16. Mubeen, S.; Habibullah, G. An integral representation of some *k*-hypergeometric functions. *Int. Math. Forum.* **2012**, *7*, 203–207.
17. Mubeen, S. *k*-analogue of Kummer's first formula. *J. Inequalities Spec. Funct.* **2012**, *3*, 41–44.
18. Korkmaz-Duzgun, D.; Erkus-Duman, E. Generating functions for *k*-hypergeometric functions. *Int. J. Appl. Phys. Math.* **2019**, *9*, 119–126.
19. Chinra, S.; Kamalappan, V.; Rakha, M.A.; Rathie, A.K. On several new contiguous function relations for *k*-hypergeometric function with two parameters. *Commun. Korean Math. Soc.* **2017**, *32*, 637–651.
20. Li, S.; Dong, Y. *k*-hypergeometric series solutions to one type of non-homogeneous *k*-hypergeometric equations. *Symmetry* **2019**, *11*, 262. [CrossRef]
21. Nisar, K.S.; Qi, F.; Rahman, G.; Mubeen, S.; Arshad, M. Some inequalities involving the extended gamma function and the Kummer confluent hypergeometric *k*-function. *J. Inequalities Appl.* **2018**, *2018*, 1–12. [CrossRef]
22. Mubeen, S.; Rehman, A. A note on *k*-gamma function and pochhammer *k*-symbol, *J. Informatics Math. Sci.* **2014**, *6*, 93–107.
23. Mubeen, S.; Habibullah, G. *k*-fractional integrals and application. *Int. J. Contemp. Math. Sci.* **2012**, *7*, 89–94.
24. Romero, L.G.; Luque, L.L.; Dorrego, G.A.; Cerutti, R.A. On the *k*-Riemann–Liouville fractional derivative. *Int. Contemp. Math. Sci.* **2013**, *8*, 41–51. [CrossRef]
25. Rahman, G.; Nisar Mubeen, S.; Sooppy, K. On generalized *k*-fractional derivative operator. *AIMS Math.* **2020**, *5*, 1936–1945. [CrossRef]
26. Mubeen, S.; Iqbal, S.; Rahman, G. Contiguous function relations and an integral representation for Appell *k*-series. *Int. Math. Res.* **2015**, *4*, 53–63. [CrossRef]
27. Kıymaz, İ. O.; Çetinkaya, A.; Agarwal, P. A study on the *k*-generalizations of some known functions and fractional operators. *J. Inequal. Spec. Funct.* **2017**, *8*, 31–41.
28. Özarslan, M.A.; Ustaoglu, C. Incomplete Caputo fractional derivative operators. *Adv. Differ. Equ.* **2018**, *2018*, 209. [CrossRef]
29. Özarslan, M.A.; Ustaoglu, C. Some Incomplete Hypergeometric Functions and Incomplete Riemann–Liouville Fractional Integral Operators. *Mathematics* **2019**, *7*, 483. [CrossRef]
30. Srivastava, H.M.; Parmar, R.K.; Chopra, P. A class of extended fractional derivative operators and associated generating relations involving hypergeometric functions. *Axioms* **2012**, *1*, 238–258. [CrossRef]
31. Choi, J.; Agarwal, P.; Jain, S. Certain fractional integral operators and extended generalized Gauss hypergeometric functions. *Kyungpook Math. J.* **2015**, *55*, 695–703. [CrossRef]
32. Özarslan, M.A.; Kürt, C. Nonhomogeneous initial and boundary value problem for the caputo-type fractional wave equation, *Adv. Differ. Equ.* **2019**, *2019*, 199. [CrossRef]
33. Fernandez, A.; Kürt, C.; Özarslan, M.A. A naturally emerging bivariate Mittag-Leffler function and associated fractional-calculus operators. *Comput. Appl. Math.* **2020**, *39*, 1–27. [CrossRef]
34. Azam, M.; Farid, G.; Rehman, M. Study of generalized type *k*-fractional derivatives. *Adv. Differ. Equ.* **2017**, *2017*, 249. [CrossRef]

© 2020 by the authors. Licensee MDPI, Basel, Switzerland. This article is an open access article distributed under the terms and conditions of the Creative Commons Attribution (CC BY) license (http://creativecommons.org/licenses/by/4.0/).

 *fractal and fractional*

Article

Electrical Circuits RC, LC, and RLC under Generalized Type Non-Local Singular Fractional Operator

Bahar Acay [1,*] and Mustafa Inc [1,2]

1. Department of Mathematics, Science Faculty, Firat University, Elazig 23119, Turkey; minc@firat.edu.tr
2. Department of Medical Research, China Medical University Hospital, China Medical University, Taichung 406040, Taiwan
* Correspondence: bacay@firat.edu.tr

Abstract: The current study is of interest when performing a useful extension of a crucial physical problem through a non-local singular fractional operator. We provide solutions that include three arbitrary parameters α, ρ, and γ for the Resistance-Capacitance (RC), Inductance-Capacitance (LC), and Resistance-Inductance-Capacitance (RLC) electric circuits utilizing a generalized type fractional operator in the sense of Caputo, called non-local M-derivative. Additionally, to keep the dimensionality of the physical parameter in the proposed model, we use an auxiliary parameter. Owing to the fact that all solutions depend on three parameters unlike the other solutions containing one or two parameters in the literature, the solutions obtained in this study have more general results. On the other hand, in order to observe the advantages of the non-local M-derivative, a comprehensive comparison is carried out in the light of experimental data. We make this comparison for the RC circuit between the non-local M-derivative and Caputo derivative. It is clearly shown on graphs that the fractional M-derivative behaves closer to the experimental data thanks to the added parameters α, ρ, and γ.

Keywords: physical problems; fractional derivatives; fractional modeling; real-world problems; electrical circuits

1. Introduction

Fractional derivatives and integrals including non-integer order are the natural generalizations of the traditional counterparts. Studies of fractional calculus in recent years have attracted considerable attention due to its advantages for modeling in various areas of science and engineering. As a result of defining non-integer order derivatives by means of integral, the non-locality property is one of its major advantages. Hence, the fractional derivatives involve data about the state variable at earlier points, and so they have a memory effect, which is useful to describe and comprehend the behavior of the complex and dynamic system. Moreover, there exist various fractional derivative and integral definitions in the literature. Accordingly, one of the main difficulties encountered in the fractional calculus is choosing an appropriate definition of the fractional operator for the problem under investigation. The Riemann–Liouville (RL) and Caputo fractional operators possess an important place in understanding the essence of fractional calculus. In particular, the Caputo fractional derivative is preferred as it is a powerful mathematical tool in application. The capabilities of the non-integer order derivatives and integrals have been shown in several rigorous studies such as the tautochrone problem, diffusion equation, control theory, models in physics, economy, biology, etc. On the other hand, some authors have proposed modified or generalized type RL and Caputo operators. It should also be mentioned that many fractional operator definitions are derived from the approach in [1]: Fractional derivative of a function with respect to (wrt) another function. Katugampola in [2] introduced a generalized-type fractional operator based on the fractional derivative of a function wrt another function. Furthermore, the authors in [3] introduced a non-local singular fractional derivative and integral by utilizing the same approach. This generalized-type fractional derivative called non-local M-derivative in the sense of Caputo is defined by:

$$_M\mathcal{D}^{\alpha,\rho,\gamma}\varphi(t) = \frac{\Gamma(\gamma+1)^{n-\alpha}}{\Gamma(n-\alpha)\rho^{n-\alpha-1}}\int_a^t ((t-a)^\rho - (\xi-a)^\rho)^{n-\alpha-1} {}_M\mathbf{D}^{n,\rho,\gamma}\varphi(\xi)\frac{d\xi}{(\xi-a)^{1-\rho}}, \qquad (1)$$

where $\alpha \in \mathbb{C}$, $n = [Re(\alpha)] + 1$, $\gamma > 0$, and ${}_M\mathbf{D}^{n,\rho,\gamma}(.)$ is the local derivative as can be seen in [4]. In [3], the Laplace transform of the Caputo type fractional M-derivative we utilize to solve the proposed model is as follows:

$$\mathcal{L}_{\rho,\gamma}\{_M\mathcal{D}^{n\alpha,\rho,\gamma}\varphi(t)\}(s) = s^\alpha \mathcal{L}_{\rho,\gamma}\{\varphi(t)\} - s^{\alpha-1}\varphi(a) - s^{\alpha-2}{}_M\mathcal{D}^{\rho,\gamma}\varphi(a) \qquad (2)$$
$$- \ldots - s^{\alpha-n+1}{}_M\mathcal{D}^{(n-2)\rho,\gamma}\varphi(a) - s^{\alpha-n}{}_M\mathcal{D}^{(n-1)\rho,\gamma}\varphi(a).$$

In a similar way, in [5], the authors presented a proportional-type non-local singular fractional operator under the local proportional derivative, which is formed by using control theory. With the help of this local derivative, novel fractional operators called proportional Caputo and constant proportional Caputo was defined in [6].

The existence of several fractional derivative and integral definitions allow us to employ the most appropriate definition for the problem addressed in order to obtain more precise results. Although many of these various definitions are quite similar, their physical interpretations may differ. It is widely known that some crucial physical properties may not be observed in classical models. In other words, such dissipative impacts on the electrical components like resistance, capacitance, and inductance as ohmic friction, non-linearity, thermal memory, etc. are not taken into account by means of the traditional approach. Consequently, there exist various physical problems handled by non-local fractional operators to capture the advantages of new generation non-integer order operators. One of the most important of the above-mentioned problems is the electrical circuits model. In [7], Gomez et al. implemented the electrical circuits with respect to the non-integer order operators to reach the analytical and numerical solutions of the proposed model, including the arbitrary parameters. In addition, the fractional Resistance-Capacitance (RC) and Resistance-Inductance-Capacitance (RLC) circuits were studied by employing some kinds of fractional operators with singular or non-singular kernels in [8]. In [9], the authors introduced the circuit elements like RC, RL, and LC via a new type non-local non-singular fractional operator under experimental data obtained from an electronic laboratory at CENIDET. The same model is investigated in [10] with a detailed comparative analysis between RL and RC circuits by means of non-singular fractional derivatives. The authors in [11] analyzed the model mentioned with the help of the generalized fractional derivative introduced by Katugampola. Moreover, the authors in [12] presented the RC, LC, and RLC circuits by employing a local-based derivative, and they obtained the analytical and numerical results. Hence, motivated by all these studies, we introduce more general solutions with the help of a generalized-type non-local singular fractional operator involving three arbitrary parameters introduced by Acay et al. in [3]. For some applications and beneficial information on fractional calculus, we refer the reader to [13–25].

The structure of the present paper is constituted as follows: In Section 2, the solutions of fractional RC, LC, and RLC electrical circuits are presented with various visual results and comprehensive interpretation. Then, in Section 3, we show a comparison between two efficient operators under an experimental data with some graphs and mention the crucial conclusions of our study.

2. Fractional Electrical Circuits

In this section, we present the RC and RLC electrical circuit including constant, exponential, and periodic sources. The fractional solutions are obtained by means of the non-local singular M-derivative containing three parameters α, ρ, and γ. Hence, we get the generalized version of the solutions obtained in the literature. The main purpose is to perform an extension of the ordinary differential equations to the fractional version via a non-local singular generalized derivative. On the other hand, preserving the physical dimensionality of the non-integer order operator is crucial in the application. In pure mathematics, generally, the integer-order derivative is replaced with non-integer order ones but this is not enough for physical problems and some applications in engineering. Therefore, dimensional modification is required for the fractional case. For this purpose, we employ the auxiliary parameter σ for the non-local fractional M-derivative in the sense of Caputo as follows:

$$\frac{d}{dt} \to \sigma^{\alpha\rho-1} {}_M\mathcal{D}^{\alpha,\rho,\gamma}, \qquad (3)$$

and

$$\frac{d^2}{dt^2} \to \sigma^{2(\alpha\rho-1)} {}_M\mathcal{D}^{\alpha,\rho,\gamma}, \qquad (4)$$

where α, ρ, and γ are arbitrary parameters, and the dimensionality of σ is the second (s). Hence, we employ this approach in order to get the solutions of the fractional electrical circuits with the help of the Caputo-type M-derivative [9,19,26].

2.1. Fractional RC Electrical Circuits under Non-Local M-Derivative in the Sense of Caputo

The RC series circuit differential equation under Kirchhoff's law can be expressed by the non-local M-derivative in the sense of Caputo considering the relations Equations (3) and (4) as below:

$$\sigma^{\alpha\rho-1} {}_M\mathcal{D}^{\alpha,\rho,\gamma} \mathbf{V_c}(t) + \frac{1}{\omega}\mathbf{V_c}(t) = \frac{1}{\omega}\mathbf{e}(t), \qquad (5)$$

where $\omega = RC$ is the time constant, R represents the resistance, C symbolizes the capacitance, the voltage on the capacitor is expressed by the function $\mathbf{V_c}(t)$, and $\mathbf{e}(t)$ is the source voltage. On the other hand, while $\sigma^{1-\alpha\rho}/RC$ is fractional time constant, $1/RC$ is a traditional time constant. It should be noted that normally the dimension of the non-local M-derivative operator is $(time)^{-\alpha\rho}$ (the parameter γ does not affect the dimension), but under favor of the term $\sigma^{\alpha\rho-1}$, we eliminate the dimension mismatch physically.

Now, let us solve the Equation (5) with the help of the Laplace transform of the non-local fractional M-derivative under three main case with different types of sources as follows:

Case 1. (Constant source). If we consider $\mathbf{e}(t) = \mathbf{e_0}$, $\mathbf{V_c}(0) = \mathbf{V_0}$ ($V_0 > 0$), we can rearrange Equation (5) as:

$${}_M\mathcal{D}^{\alpha,\rho,\gamma} \mathbf{V_c}(t) + \frac{\sigma^{1-\alpha\rho}}{\omega}\mathbf{V_c}(t) = \frac{\sigma^{1-\alpha\rho}}{\omega}\mathbf{e_0}. \qquad (6)$$

Applying LT of the non-local M-derivative in the Caputo sense, we have:

$$\mathcal{L}_{\rho,\gamma}\{{}_M\mathcal{D}^{\alpha,\rho,\gamma}\mathbf{V_c}(t)\} + \frac{\sigma^{1-\alpha\rho}}{\omega}\mathcal{L}_{\rho,\gamma}\{\mathbf{V_c}(t)\} = \mathcal{L}_{\rho,\gamma}\left\{\frac{\sigma^{1-\alpha\rho}}{\omega}\mathbf{e_0}\right\}, \qquad (7)$$

$$s^\alpha \mathcal{L}_{\rho,\gamma}\{\mathbf{V_c}(t)\} - s^{\alpha-1}\mathbf{V_c}(0) + \frac{\sigma^{1-\alpha\rho}}{\omega}\mathcal{L}_{\rho,\gamma}\{\mathbf{V_c}(t)\} = \frac{\mathbf{e_0}\sigma^{1-\alpha\rho}}{s\omega}, \qquad (8)$$

after some arrangements, one can get:

$$\mathcal{L}_{\rho,\gamma}\{\mathbf{V_c}(t)\} = \mathbf{V_0}\frac{s^{\alpha-1}}{s^\alpha + \frac{\sigma^{1-\alpha\rho}}{\omega}} + \frac{\mathbf{e_0}\sigma^{1-\alpha\rho}}{\omega}\frac{1}{s\left(s^\alpha + \frac{\sigma^{1-\alpha\rho}}{\omega}\right)}, \qquad (9)$$

hence if we take the inverse LT of Equation (9), we reach the following solution:

$$\mathbf{V_c}(t) = \mathbf{V_0}\mathbf{E_\alpha}\left(-\frac{\sigma^{1-\alpha\rho}}{\omega}\left(\Gamma(\gamma+1)\frac{t^\rho}{\rho}\right)^\alpha\right) + \mathbf{e_0}\left[1 - \mathbf{E_\alpha}\left(-\frac{\sigma^{1-\alpha\rho}}{\omega}\left(\Gamma(\gamma+1)\frac{t^\rho}{\rho}\right)^\alpha\right)\right]. \qquad (10)$$

where $\mathbf{E}_\alpha(.)$ is the Mittag–Leffler function.

It can be seen that the fractional solution follows exponential dynamics if α, ρ, and γ are closer to 1. In Figures 1 and 2 we show the plot for Case 1 (constant source) when $R = 1$ Ω, $C = 10$ F, $\mathbf{e}_0 = 5$ V, and $\mathbf{V}_c(0) = 10$ V, and in Figures 3 and 4, we use the values $R = 1$ Ω, $C = 10$ F, $\mathbf{e}_0 = 5$ V, and $\mathbf{V}_c(0) = 0$ V. We observe the behavior of voltage across the capacitor in the RC circuit for $\mathbf{e}(t) = \mathbf{e}_0$ in Figures 1–4 when α changes; $\alpha = 1, 0.9, 0.8, 0.7$, $\rho = 0.9$, and $\gamma = 0.9$, when ρ changes; $\rho = 1, 0.9, 0.8, 0.7$, $\alpha = 0.9$, and $\gamma = 1.5$, and when γ changes; $\gamma = 1, 1, 6, 1.8, 2$, $\alpha = 0.9$, and $\rho = 0.9$. In this way, the impact of the parameters α, ρ, and γ can clearly by observed on the solutions curves separately. On the other hand, in Figure 1, we observe that for the small values of α, the solution curve tends to stabilize in less time with exponential behavior. However for a classical case (when $\alpha = 1$, $\rho = 1$, and $\gamma = 1$) it stabilizes in longer time. In Figure 2a, we see similar behavior in the solutions curves when ρ changes. However, the effect of the parameter γ is different from the effect of the parameters α and ρ as can be seen in Figures 2b and 4b. It can be observed that for smaller values of γ, the solution curve approaches to stabilize in a longer time. Moreover, in Figures 1 and 2, one can see that the solution curves are exponentially decreasing, and in Figures 3 and 4, the exponentially increasing overdamped system.

Case 2. (Exponential source). Let $\mathbf{e}(t) = \mathbf{e}_0 e^{-\lambda\Gamma(\gamma+1)\frac{t^\rho}{\rho}}$, $\mathbf{V}_c(0) = V_0$ ($V_0 > 0$). Then we can rewrite the Equation (5) as follows:

$$_M\mathcal{D}^{\alpha,\rho,\gamma}\mathbf{V}_c(t) + \frac{\sigma^{1-\alpha\rho}}{\omega}\mathbf{V}_c(t) = \frac{\sigma^{1-\alpha\rho}}{\omega}\mathbf{e}_0 e^{-\lambda\Gamma(\gamma+1)\frac{t^\rho}{\rho}}. \tag{11}$$

If we take the LT of the Equation (11), we have:

$$\mathcal{L}_{\rho,\gamma}\{_M\mathcal{D}^{\alpha,\rho,\gamma}\mathbf{V}_c(t)\} + \frac{\sigma^{1-\alpha\rho}}{\omega}\mathcal{L}_{\rho,\gamma}\{\mathbf{V}_c(t)\} = \frac{\mathbf{e}_0\sigma^{1-\alpha\rho}}{\omega}\mathcal{L}_{\rho,\gamma}\left\{e^{-\lambda\Gamma(\gamma+1)\frac{t^\rho}{\rho}}\right\}, \tag{12}$$

$$s^\alpha \mathcal{L}_{\rho,\gamma}\{\mathbf{V}_c(t)\} - s^{\alpha-1}\mathbf{V}_c(0) + \frac{\sigma^{1-\alpha\rho}}{\omega}\mathcal{L}_{\rho,\gamma}\{\mathbf{V}_c(t)\} = \frac{\mathbf{e}_0\sigma^{1-\alpha\rho}}{\omega(s+\lambda)}, \tag{13}$$

$$\mathcal{L}_{\rho,\gamma}\{\mathbf{V}_c(t)\} = \frac{\mathbf{e}_0\sigma^{1-\alpha\rho}}{\omega}\frac{1}{(s+\lambda)\left(s^\alpha + \frac{\sigma^{1-\alpha\rho}}{\omega}\right)} + V_0\frac{s^\alpha}{s\left(s^\alpha + \frac{\sigma^{1-\alpha\rho}}{\omega}\right)}, \tag{14}$$

applying the inverse LT and the convolution theorem, we can obtain the solution as:

$$\begin{aligned}\mathbf{V}_c(t) &= V_0 E_\alpha\left(-\frac{\sigma^{1-\alpha\rho}}{\omega}\left(\Gamma(\gamma+1)\frac{t^\rho}{\rho}\right)^\alpha\right) \\ &+ \frac{\mathbf{e}_0\sigma^{1-\alpha\rho}}{\omega}\Gamma(\gamma+1)\int_0^t \left(\Gamma(\gamma+1)\frac{t^\rho}{\rho} - \Gamma(\gamma+1)\frac{\tau^\rho}{\rho}\right)^{\alpha-1} \\ &\times E_{\alpha,\alpha}\left(-\frac{\sigma^{1-\alpha\rho}}{\omega}\left(\Gamma(\gamma+1)\frac{t^\rho}{\rho} - \Gamma(\gamma+1)\frac{\tau^\rho}{\rho}\right)^\alpha\right) \\ &\times \exp\left(-\lambda\Gamma(\gamma+1)\frac{\tau^\rho}{\rho}\right)\tau^{\rho-1}d\tau.\end{aligned} \tag{15}$$

Case 3. (Oscillatory source). If we suppose that $\mathbf{e}(t) = \mathbf{e}_0 \cos\left(\theta\Gamma(\gamma+1)\frac{t^\rho}{\rho}\right)$, $\mathbf{V}_c(0) = V_0$ ($V_0 > 0$), then we can write:

$$_M\mathcal{D}^{\alpha,\rho,\gamma}\mathbf{V}_c(t) + \frac{\sigma^{1-\alpha\rho}}{\omega}\mathbf{V}_c(t) = \frac{\sigma^{1-\alpha\rho}}{\omega}\mathbf{e}_0\cos\left(\theta\Gamma(\gamma+1)\frac{t^\rho}{\rho}\right). \tag{16}$$

Taking LT of the Equation (16), we readily have:

$$\mathcal{L}_{\rho,\gamma}\{_M\mathcal{D}^{\alpha,\rho,\gamma}\mathbf{V}_c(t)\} + \frac{\sigma^{1-\alpha\rho}}{\omega}\mathcal{L}_{\rho,\gamma}\{\mathbf{V}_c(t)\} = \frac{\mathbf{e}_0\sigma^{1-\alpha\rho}}{\omega}\mathcal{L}_{\rho,\gamma}\left\{\cos\left(\theta\Gamma(\gamma+1)\frac{t^\rho}{\rho}\right)\right\}, \tag{17}$$

$$s^\alpha \mathcal{L}_{\rho,\gamma}\{V_c(t)\} - s^{\alpha-1} V_c(0) + \frac{\sigma^{1-\alpha\rho}}{\omega} \mathcal{L}_{\rho,\gamma}\{V_c(t)\} = \frac{e_0 \sigma^{1-\alpha\rho} s}{\omega(\theta^2 + s^2)}, \tag{18}$$

$$\mathcal{L}_{\rho,\gamma}\{V_c(t)\} = \frac{e_0 \sigma^{1-\alpha\rho}}{\omega} \frac{s}{(\theta^2 + s^2)\left(s^\alpha + \frac{\sigma^{1-\alpha\rho}}{\omega}\right)} + V_0 \frac{s^\alpha}{s\left(s^\alpha + \frac{\sigma^{1-\alpha\rho}}{\omega}\right)}, \tag{19}$$

after applying inverse LT transform and convolution theorem, one can reach the solution below:

$$\begin{aligned}
V_c(t) &= V_0 E_\alpha \left(-\frac{\sigma^{1-\alpha\rho}}{\omega}\left(\Gamma(\gamma+1)\frac{t^\rho}{\rho}\right)^\alpha\right) \\
&+ \frac{e_0 \sigma^{1-\alpha\rho}}{\omega}\Gamma(\gamma+1) \int_0^t \left(\Gamma(\gamma+1)\frac{t^\rho}{\rho} - \Gamma(\gamma+1)\frac{\tau^\rho}{\rho}\right)^{\alpha-1} \\
&\times E_{\alpha,\alpha}\left(-\frac{\sigma^{1-\alpha\rho}}{\omega}\left(\Gamma(\gamma+1)\frac{t^\rho}{\rho} - \Gamma(\gamma+1)\frac{\tau^\rho}{\rho}\right)^\alpha\right) \\
&\times \cos\left(\theta \Gamma(\gamma+1)\frac{t^\rho}{\rho}\right) \tau^{\rho-1} d\tau.
\end{aligned} \tag{20}$$

For oscillatory source case involving the angular frequency θ ($e(t) = e_0 \cos(\theta t)$), we present Figures 5–8 for various values of the α, γ, and ρ when $R = 1\,\Omega$, $C = 10\,F$, $e_0 = 10\,V$, $\theta = 60\,Hz$, and $V_c(0) = 10\,V$. We should note that in the case of standard approach $\alpha = 1$, $\rho = 1$, and $\gamma = 1$, some losses which are based on the ohmic friction, temperature, and so on are not considered. However, the non-integer order approach enables us to examine the proposed physical problem more precisely. It is also seen that the solutions curves with different values of α, ρ, and γ are below or under the traditional solution curve for a time. This situation varies for different arbitrary parameter values. On the other hand, we can see the behavior of voltage across the capacitor in the RC circuit for $e(t) = e_0 \cos\left(\Gamma(\gamma+1)\frac{t^\rho}{\rho}\right)$ in Figures 5–8 when α changes; $\alpha = 1, 0.995, 0.99, 0.985$, $\rho = 0.9$, and $\gamma = 1.5$, when ρ changes; $\rho = 1, 0.9, 0.8, 0.7$, $\alpha = 0.9$, and $\gamma = 0.9$, and when γ changes; $\gamma = 1, 1, 15, 1.2, 1.25$, $\alpha = 0.9$, and $\rho = 0.9$. Furthermore, in Figures 5 and 6, we can also observe that the period may change under the fractional-order derivative, the apparent motion of the solution curves obtained by employing the arbitrary order may appear more complicated, and the extremes of the solution function can change for different values of fractional order. In addition, for different values R and C, the wave height and length change in Figures 7 and 8, respectively.

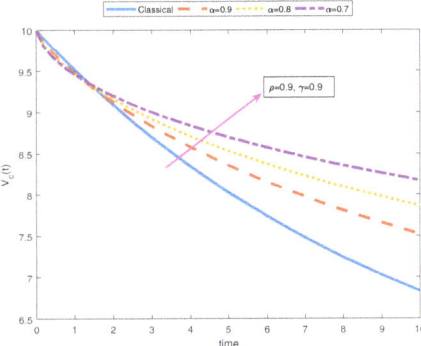

Figure 1. This figure corresponds to the function $V_c(t)$ with constant source for different values of α, ρ, and γ in order to show the effect of α on solution curves when $R = 1\,\Omega$, $C = 10\,F$, $e_0 = 5\,V$, and $V_c(0) = 10$.

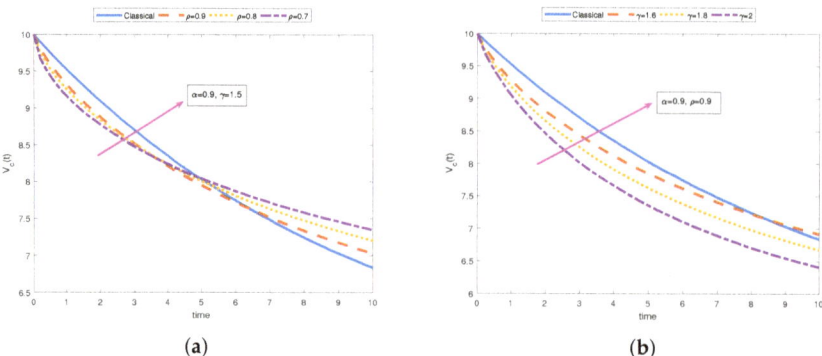

Figure 2. The figure (**a**) corresponds to the function $\mathbf{V}_c(t)$ with constant source for different values of α, ρ, and γ in order to show the effect of α and γ, and also the figure (**b**) is plotted to show the effect of α and ρ when $R = 1\ \Omega$, $C = 10$ F, $\mathbf{e}_0 = 5$ V, and $\mathbf{V}_c(0) = 10$.

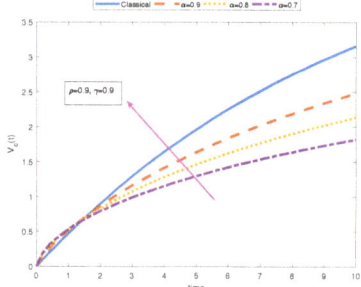

Figure 3. This figure corresponds to the function $\mathbf{V}_c(t)$ with a constant source for different values of α, ρ, and γ in order to show the effect of α on solution curves when $R = 1\ \Omega$, $C = 10$ F, $\mathbf{e}_0 = 5$ V, and $\mathbf{V}_c(0) = 0$.

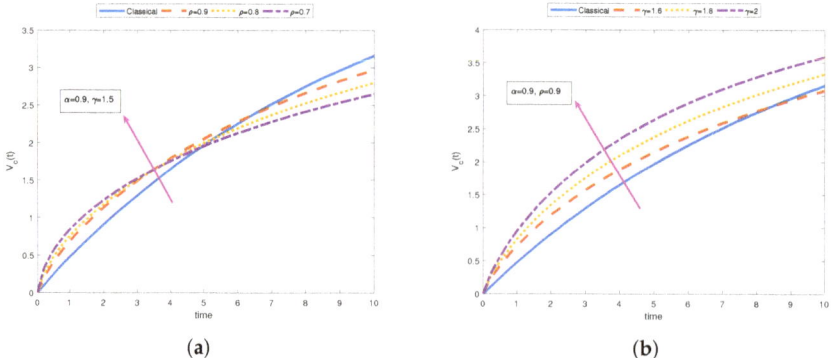

Figure 4. The figure (**a**) corresponds to the function $\mathbf{V}_c(t)$ with constant source for different values of α, ρ, and γ in order to show the effect of α and γ, and also the figure (**b**) is plotted to show the effect of α and ρ when $R = 1\ \Omega$, $C = 10$ F, $\mathbf{e}_0 = 5$ V, and $\mathbf{V}_c(0) = 0$.

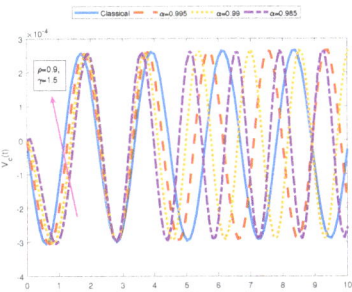

Figure 5. This figure corresponds to the function $V_c(t)$ with an oscillatory source for different values of α, ρ, and γ in order to show the effect of α on solution curves when $R = 1\,\Omega$, $C = 10$ F, $\mathbf{e}_0 = 10$ V, and $\theta = 60$ Hz, $\mathbf{V}_c(0) = 10$.

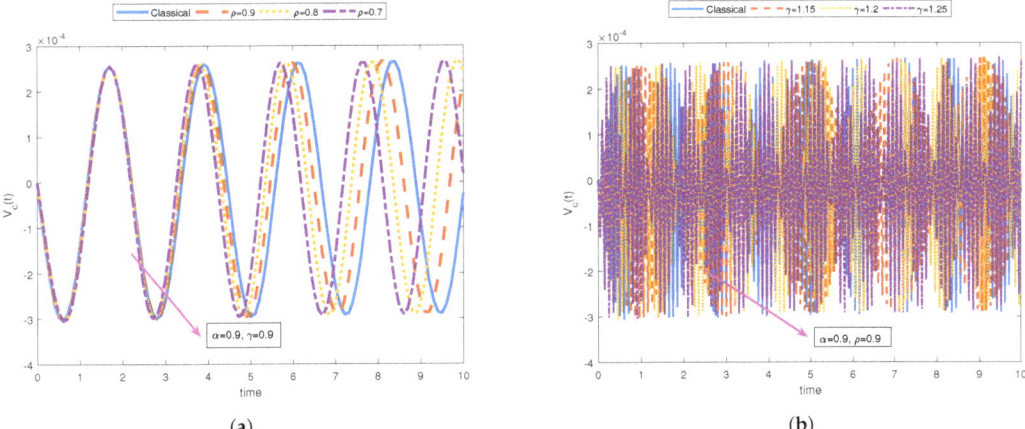

Figure 6. The figure (**a**) corresponds to the function $V_c(t)$ with an oscillatory source for different values of α, ρ, and γ in order to show the effect of α and γ, and also the figure (**b**) is plotted to show the effect of α and ρ on solution curves when $R = 1\,\Omega$, $C = 10$ F, $\mathbf{e}_0 = 10$ V, and $\theta = 60$ Hz, $\mathbf{V}_c(0) = 10$.

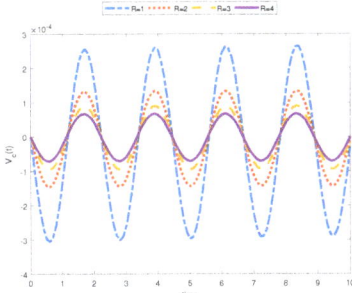

Figure 7. This figure corresponds to the function $V_c(t)$ with an oscillatory source for some values of α, ρ, and γ in order to show the effect of resistance on solution curves when $\alpha = 1$, $\rho = 1$, $\gamma = 1$, $C = 10$ F, $\mathbf{e}_0 = 10$ V, and $\theta = 60$ Hz, $\mathbf{V}_c(0) = 10$.

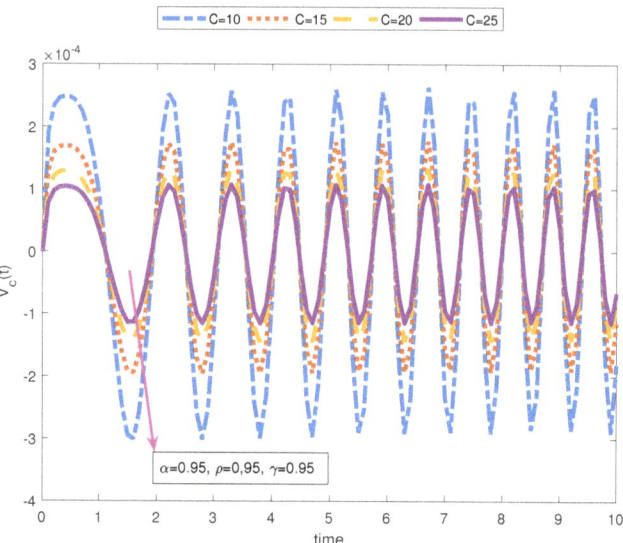

Figure 8. This figure corresponds to the function $\mathbf{V}_c(t)$ with an oscillatory source for some values of α, ρ, and γ in order to show the effect of capacitance on solution curves when $\alpha = 0.95$, $\rho = 0.95$, $\gamma = 1.5$, $C = 10$ F, $\mathbf{e}_0 = 10$ V, and $\theta = 60$ Hz, $\mathbf{V}_c(0) = 10$.

2.2. Fractional Inductance-Capacitance (LC) Electrical Circuits under Non-Local M-Derivative in the Sense of Caputo

If the law of Kirchhoff is applied, then the LC series circuit differential equation under the relations Equations (3) and (4) can be presented by:

$$\sigma^{2(\alpha\rho-1)}{}_M\mathcal{D}^{2\alpha,\rho,\gamma}\mathbf{I}(t) + \frac{1}{\eta}\mathbf{I}(t) = \frac{C}{\eta}\mathbf{e}(t), \tag{21}$$

where $\eta = LC$, L represents the inductance, the capacitance is denoted by C, and $\mathbf{e}(t)$ stands for source voltage. Now, we solve the above-stated equation by employing LT of the non-local fractional M-derivative with three case including different sources as below:

Case 1. (Constant source). Supposing $\mathbf{e}(t) = \mathbf{e}_0$, $\mathbf{I}(0) = \mathbf{I}_0$ ($\mathbf{I}_0 > 0$), ${}_M\mathbf{D}^{æ,fl}\mathbf{I}(0) = 0$, we can express the Equation (21) as:

$$_M\mathcal{D}^{2\alpha,\rho,\gamma}\mathbf{I}(t) + \frac{\sigma^{2(1-\alpha\rho)}}{\eta}\mathbf{I}(t) = \frac{C\sigma^{2(1-\alpha\rho)}}{\eta}\mathbf{e}_0, \tag{22}$$

where $\eta = LC$ and C is the capacitance. If we apply the LT to Equation (22), we have:

$$\mathcal{L}_{\rho,\gamma}\{{}_M\mathcal{D}^{2\alpha,\rho,\gamma}\mathbf{I}(t)\} + \frac{\sigma^{2(1-\alpha\rho)}}{\eta}\mathcal{L}_{\rho,\gamma}\{\mathbf{I}(t)\} = \mathcal{L}_{\rho,\gamma}\left\{\frac{C\sigma^{2(1-\alpha\rho)}\mathbf{e}_0}{\eta}\right\}, \tag{23}$$

$$s^\alpha \mathcal{L}_{\rho,\gamma}\{\mathbf{I}(t)\} - s^{\alpha-1}\mathbf{I}(0) - s^{\alpha-2}{}_M\mathbf{D}^{æ,fl}\mathbf{I}(0) + \frac{\sigma^{2(1-\alpha\rho)}}{\eta}\mathcal{L}_{\rho,\gamma}\{\mathbf{I}(t)\} = \frac{C\sigma^{2(1-\alpha\rho)}\mathbf{e}_0}{s\eta}, \tag{24}$$

$$\mathcal{L}_{\rho,\gamma}\{\mathbf{I}(t)\} = \mathbf{I}_0 \frac{s^{\alpha-1}}{s^\alpha + \frac{\sigma^{2(1-\alpha\rho)}}{\eta}} + \frac{C\sigma^{2(1-\alpha\rho)}\mathbf{e}_0}{\eta} \frac{1}{s\left(s^\alpha + \frac{\sigma^{2(1-\alpha\rho)}}{\eta}\right)}, \tag{25}$$

and if we apply the inverse LT to the Equation (25), we reach the following solution:

$$\mathbf{I}(t) = \mathbf{I}_0 E_\alpha\left(-\frac{\sigma^{2(1-\alpha\rho)}}{\eta}\left(\Gamma(\gamma+1)\frac{t^\rho}{\rho}\right)^\alpha\right) + Ce_0\left[1 - E_\alpha\left(-\frac{\sigma^{2(1-\alpha\rho)}}{\eta}\left(\Gamma(\gamma+1)\frac{t^\rho}{\rho}\right)^\alpha\right)\right]. \tag{26}$$

$\frac{1}{\eta} = \frac{1}{LC}$ in Equation (21) represents the natural angular frequency, and the initial charge of the capacitor is denoted by $\mathbf{I}_0$. In Figure 9, A and B correspond to Equation (26) with constant source when $\mathbf{I}(0) = 0$ and $\mathbf{I}(0) = 10$, respectively. These plots are obtained when $\alpha = 1, 0.9, 0.8, 0.7$, $\rho = 1, 0.9, 0.8, 0.7$, and $\gamma = 0.7$. The system is exponentially increasing in A and exponentially decreasing in B. We see that the solution curve tends faster to the steady state for small values of α and ρ.

Case 2. (Exponential source). Let $e(t) = e_0 e^{-\lambda\Gamma(\gamma+1)\frac{t^\rho}{\rho}}$, $\mathbf{I}(0) = \mathbf{I}_0$ ($\mathbf{I}_0 > 0$), ${}_M D^{\rho,\gamma}\mathbf{I}(0) = 0$, then Equation (21) can be written as follows:

$$_M\mathcal{D}^{2\alpha,\rho,\gamma}\mathbf{I}(t) + \frac{\sigma^{2(1-\alpha\rho)}}{\eta}\mathbf{I}(t) = \frac{C\sigma^{2(1-\alpha\rho)}}{\eta}e_0 e^{-\lambda\Gamma(\gamma+1)\frac{t^\rho}{\rho}}, \tag{27}$$

applying the LT to Equation (27), we have:

$$\mathcal{L}_{\rho,\gamma}\{{}_M D^{2\alpha,\rho,\gamma}\mathbf{I}(t)\} + \frac{\sigma^{2(1-\alpha\rho)}}{\eta}\mathcal{L}_{\rho,\gamma}\{\mathbf{I}(t)\} = \frac{C\sigma^{2(1-\alpha\rho)}e_0}{\eta}\mathcal{L}_{\rho,\gamma}\{e^{-\lambda\Gamma(\gamma+1)\frac{t^\rho}{\rho}}\}, \tag{28}$$

$$s^\alpha \mathcal{L}_{\rho,\gamma}\{\mathbf{I}(t)\} - s^{\alpha-1}\mathbf{I}(0) - s^{\alpha-2}{}_M D^{\alpha,\rho,\gamma}\mathbf{I}(0) + \frac{\sigma^{2(1-\alpha\rho)}}{\eta}\mathcal{L}_{\rho,\gamma}\{\mathbf{I}(t)\} = \frac{C\sigma^{2(1-\alpha\rho)}e_0}{\eta(s+\lambda)}, \tag{29}$$

$$\mathcal{L}_{\rho,\gamma}\{\mathbf{I}(t)\} = \mathbf{I}_0 \frac{s^{\alpha-1}}{s^\alpha + \frac{\sigma^{2(1-\alpha\rho)}}{\eta}} + \frac{C\sigma^{2(1-\alpha\rho)}e_0}{\eta}\frac{1}{(s+\lambda)\left(s^\alpha + \frac{\sigma^{2(1-\alpha\rho)}}{\eta}\right)}, \tag{30}$$

after taking inverse LT, one can attain the following solution:

$$\begin{aligned}\mathbf{I}(t) &= \mathbf{I}_0 E_\alpha\left(-\frac{\sigma^{2(1-\alpha\rho)}}{\eta}\left(\Gamma(\gamma+1)\frac{t^\rho}{\rho}\right)^\alpha\right) \\ &+ \frac{C\sigma^{2(1-\alpha\rho)}e_0}{\eta}\Gamma(\gamma+1)\int_0^t\left(\Gamma(\gamma+1)\frac{t^\rho}{\rho} - \Gamma(\gamma+1)\frac{\tau^\rho}{\rho}\right)^{\alpha-1} \\ &\times E_{\alpha,\alpha}\left(-\frac{\sigma^{2(1-\alpha\rho)}}{\eta}\left(\Gamma(\gamma+1)\frac{t^\rho}{\rho} - \Gamma(\gamma+1)\frac{\tau^\rho}{\rho}\right)^\alpha\right) \\ &\times \exp\left(-\lambda\Gamma(\gamma+1)\frac{\tau^\rho}{\rho}\right)\tau^{\rho-1}d\tau.\end{aligned} \tag{31}$$

Figure 10 corresponds to solution Equation (31) including exponential source when $\alpha = 1, 0.9, 0.8, 0.7$, $\rho = 1, 0.9, 0.8, 0.7$, $\gamma = 0.9$, $C = 0.5$ F, $L = 2.4$ H, $\lambda = 0.05$, $e_0 = 5$ V, and $\mathbf{I}(0) = 0$. From Figures 10 and 11, we observe an oscillatory behavior and the effect of the parameters α, ρ, and γ on the function $\mathbf{I}(t)$. Furthermore, in Figure 11 corresponding to $\mathbf{I}(t)$ with exponential source, the impact of arbitrary parameters when $\alpha = 0.9$, $\rho = 0.9$, $\gamma = 0.9$ in A, and $\alpha = 0.5$, $\rho = 0.5$, $\gamma = 0.5$ in B for $L = 2.4, 3.4, 4.4, 5.4$ can clearly be seen. It is worth mentioning that in Figures 9 and 10, having exponential behavior, while the classical solution function tends to stabilize slower, the fractional solution function with smaller values of α and ρ approach to stabilize in less time. On the other hand, one can see the oscillatory behavior underdamped system in Figures 10 and 11. In Figure 12, the different values of the parameter L change the wave height critically.

Case 3. (Oscillatory source). Assuming that $\mathbf{e}(t) = \mathbf{e}_0 \cos\left(\theta\Gamma(\gamma+1)\frac{t^\rho}{\rho}\right)$, $\mathbf{I}(0) = \mathbf{I}_0$ ($\mathbf{I}_0 > 0$), and ${}_M\mathbf{D}^{æ,\text{fl}}\mathbf{I}(0) = 0$, we present the Equation (21) in the following form:

$$ {}_M\mathcal{D}^{2\alpha,\rho,\gamma}\mathbf{I}(t) + \frac{\sigma^{2(1-\alpha\rho)}}{\eta}\mathbf{I}(t) = \frac{C\sigma^{2(1-\alpha\rho)}}{\eta}\mathbf{e}_0 \cos\left(\theta\Gamma(\gamma+1)\frac{t^\rho}{\rho}\right), \tag{32}$$

by applying the LT, we get:

$$\mathcal{L}_{\rho,\gamma}\{{}_M\mathcal{D}^{2\alpha,\rho,\gamma}\mathbf{I}(t)\} + \frac{\sigma^{2(1-\alpha\rho)}}{\eta}\mathcal{L}_{\rho,\gamma}\{\mathbf{I}(t)\} = \frac{C\sigma^{2(1-\alpha\rho)}\mathbf{e}_0}{\eta}\mathcal{L}_{\rho,\gamma}\left\{\cos\left(\theta\Gamma(\gamma+1)\frac{t^\rho}{\rho}\right)\right\}, \tag{33}$$

$$s^\alpha \mathcal{L}_{\rho,\gamma}\{\mathbf{I}(t)\} - s^{\alpha-1}\mathbf{I}(0) - s^{\alpha-2}{}_M\mathbf{D}^{æ,\text{fl}}\mathbf{I}(0) + \frac{\sigma^{2(1-\alpha\rho)}}{\eta}\mathcal{L}_{\rho,\gamma}\{\mathbf{I}(t)\} = \frac{Cs\sigma^{2(1-\alpha\rho)}\mathbf{e}_0}{\eta(\theta^2+s^2)}, \tag{34}$$

$$\mathcal{L}_{\rho,\gamma}\{\mathbf{I}(t)\} = \mathbf{I}_0 \frac{s^{\alpha-1}}{s^\alpha + \frac{\sigma^{2(1-\alpha\rho)}}{\eta}} + \frac{C\sigma^{2(1-\alpha\rho)}\mathbf{e}_0}{\eta}\frac{s}{(\theta^2+s^2)(s^\alpha + \frac{\sigma^{2(1-\alpha\rho)}}{\eta})}, \tag{35}$$

and if we apply the inverse LT, then we get the solution below:

$$\begin{aligned}\mathbf{I}(t) &= \mathbf{I}_0 \mathbf{E}_\alpha\left(-\frac{\sigma^{2(1-\alpha\rho)}}{\eta}\left(\Gamma(\gamma+1)\frac{t^\rho}{\rho}\right)^\alpha\right) \\ &+ \frac{C\sigma^{2(1-\alpha\rho)}\mathbf{e}_0}{\eta}\Gamma(\gamma+1)\int_0^t \left(\Gamma(\gamma+1)\frac{t^\rho}{\rho} - \Gamma(\gamma+1)\frac{\tau^\rho}{\rho}\right)^{\alpha-1} \\ &\times \mathbf{E}_{\alpha,\alpha}\left(-\frac{\sigma^{2(1-\alpha\rho)}}{\eta}\left(\Gamma(\gamma+1)\frac{t^\rho}{\rho} - \Gamma(\gamma+1)\frac{\tau^\rho}{\rho}\right)^\alpha\right) \\ &\times \cos\left(\theta\Gamma(\gamma+1)\frac{t^\rho}{\rho}\right)\tau^{\rho-1}d\tau.\end{aligned} \tag{36}$$

Figure 12 is plotted for the solution Equation (36) when $L = 2, 3, 4, 5$ and $\gamma = 0.95$.

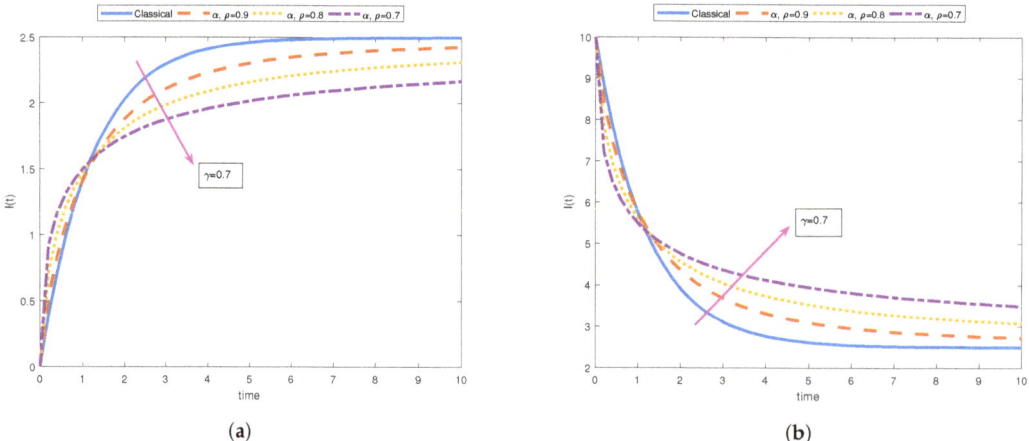

Figure 9. The figure (**a**) corresponds to the function $\mathbf{I}(t)$ with constant source for various values of α, ρ, and γ in order to show the effect of arbitrary parameter γ when $\gamma = 0.7$, $C = 0.5$ F, $L = 2.4$, $\mathbf{e}_0 = 5$ V, $\mathbf{I}(0) = 0$, and similarly the figure (**b**) is plotted to show the effect of γ on solution curves when $\gamma = 0.7$, $C = 0.5$ F, $L = 2.4$, $\mathbf{e}_0 = 5$ V, and $\mathbf{I}(0) = 10$.

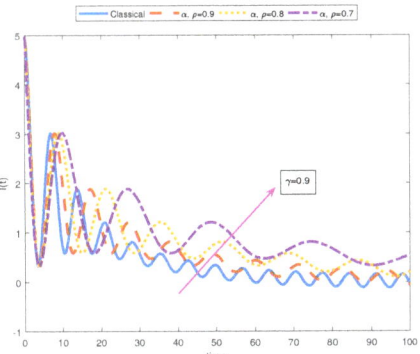

Figure 10. This graph corresponds to the solution Equation (31) containing exponential source for various values of α and ρ when $\gamma = 0.9$, $C = 0.5$ F, $L = 2.4$ H, $\lambda = 0.05$, $\mathbf{e}_0 = 5$ V, and $\mathbf{I}(0) = 0$.

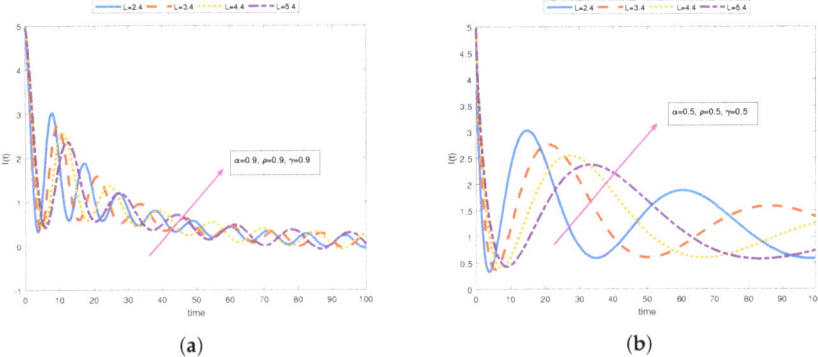

Figure 11. The figure (**a**) corresponds to the solution Equation (31) containing exponential source for various values of α, ρ, and γ in order to see the effect of parameter L when $C = 0.5$ F, $\lambda = 0.05$, $\mathbf{e}_0 = 5$ V, and $\mathbf{I}(0) = 0$, and similarly the figure (**b**) is plotted to show the effect of L on solution curves when $C = 0.5$ F, $\lambda = 0.05$, $\mathbf{e}_0 = 5$ V, and $\mathbf{I}(0) = 0$.

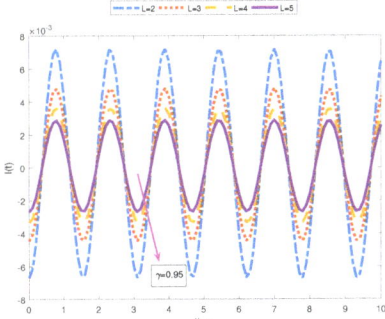

Figure 12. This graph corresponds to solution Equation (36) including oscillatory source for various values of α and ρ and γ in order to see the effect of L when $C = 47$ F, $\mathbf{e}_0 = 50$ V, $\theta = 60$ Hz, and $\mathbf{I}(0) = 0$.

2.3. Fractional RLC Electrical Circuits under Non-Local M-Derivative in the Sense of Caputo

The fractional RLC series circuit differential equation can be presented according to the non-local M-derivative as below:

$$\sigma^{2(\alpha\rho-1)}{}_M D^{2\alpha,\rho,\gamma}\mathbf{q}(t) + \frac{CR}{\delta}\sigma^{\alpha\rho-1}{}_M D^{\alpha,\rho,\gamma}\mathbf{q}(t) + \frac{1}{\delta}\mathbf{q}(t) = \frac{C}{\delta}\mathbf{e}(t), \tag{37}$$

where $\delta = LC$, L denotes the inductance, the capacitance is represented by C, R stands for the resistance, and $\mathbf{e}(t)$ is the source voltage. Let us solve Equation (37) under the fractional non-local M-derivative with the help of the LT. We present three cases including different types of sources as follows:

Case 1. (Constant source). Assuming that $\mathbf{e}(t) = \mathbf{e}_0$, $\mathbf{q}(0) = \mathbf{q}_0$, and ${}_M D^{\rho,\gamma}\mathbf{q}(0) = 0$, we give Equation (37) the following form:

$$_M D^{2\alpha,\rho,\gamma}\mathbf{q}(t) + \frac{CR}{\delta}\sigma^{1-\alpha\rho}{}_M D^{\alpha,\rho,\gamma}\mathbf{q}(t) + \frac{\sigma^{2(1-\alpha\rho)}}{\delta}\mathbf{q}(t) = \frac{C\sigma^{2(1-\alpha\rho)}\mathbf{e}_0}{\delta}, \tag{38}$$

by applying the LT of the Equation (38), then we attain:

$$\mathcal{L}_{\rho,\gamma}\{{}_M D^{2\alpha,\rho,\gamma}\mathbf{q}(t)\} + \frac{CR\sigma^{1-\alpha\rho}}{\delta}\mathcal{L}_{\rho,\gamma}\{{}_M D^{\alpha,\rho,\gamma}\mathbf{q}(t)\} + \frac{\sigma^{2(1-\alpha\rho)}}{\delta}\mathcal{L}_{\rho,\gamma}\{\mathbf{q}(t)\} = \mathcal{L}_{\rho,\gamma}\left\{\frac{C\sigma^{2(1-\alpha\rho)}\mathbf{e}_0}{\delta}\right\}, \tag{39}$$

$$\begin{aligned}
& s^\alpha\left[\mathcal{L}_{\rho,\gamma}\{\mathbf{q}(t)\} - s^{\alpha-1}\mathbf{q}(0) - s^{\alpha-2}{}_M D^{\rho,\gamma}\mathbf{q}(0)\right] \\
& + \frac{CR\sigma^{1-\alpha\rho}}{\delta}\left[s^\alpha \mathcal{L}_{\rho,\gamma}\{\mathbf{q}(t)\} - s^{\alpha-1}\mathbf{q}_0\right] + \frac{\sigma^{2(1-\alpha\rho)}}{\delta}\mathcal{L}_{\rho,\gamma}\{\mathbf{q}(t)\} \\
& = \frac{C\sigma^{2(1-\alpha\rho)}\mathbf{e}_0}{s\delta},
\end{aligned} \tag{40}$$

$$\begin{aligned}
\mathcal{L}_{\rho,\gamma}\{\mathbf{q}(t)\} & = \mathbf{q}_0\frac{s^{\alpha-1}}{s^\alpha\left(1 + \frac{CR\sigma^{1-\alpha\rho}}{\delta}\right) + \frac{\sigma^{2(1-\alpha\rho)}}{\delta}} + \mathbf{q}_0\frac{CR\sigma^{1-\alpha\rho}}{\delta}\frac{s^{\alpha-1}}{s^\alpha\left(1 + \frac{CR\sigma^{1-\alpha\rho}}{\delta}\right) + \frac{\sigma^{2(1-\alpha\rho)}}{\delta}} \\
& + \frac{C\sigma^{2(1-\alpha\rho)}\mathbf{e}_0}{\delta}\frac{1}{s\left[s^\alpha\left(1 + \frac{CR\sigma^{1-\alpha\rho}}{\delta}\right) + \frac{\sigma^{2(1-\alpha\rho)}}{\delta}\right]},
\end{aligned} \tag{41}$$

$$\begin{aligned}
\mathcal{L}_{\rho,\gamma}\{\mathbf{q}(t)\} & = \mathbf{q}_0\frac{\delta}{\delta + CR\sigma^{1-\alpha\rho}}\frac{s^{\alpha-1}}{s^\alpha + \frac{\sigma^{2(1-\alpha\rho)}}{\delta+CR\sigma^{1-\alpha\rho}}} + \mathbf{q}_0\frac{CR\sigma^{1-\alpha\rho}}{\delta + CR\sigma^{1-\alpha\rho}}\frac{s^{\alpha-1}}{s^\alpha + \frac{\sigma^{2(1-\alpha\rho)}}{\delta+CR\sigma^{1-\alpha\rho}}} \\
& + \frac{C\sigma^{2(1-\alpha\rho)}\mathbf{e}_0}{\delta + CR\sigma^{1-\alpha\rho}}\frac{1}{s\left(s^\alpha + \frac{\sigma^{2(1-\alpha\rho)}}{\delta+CR\sigma^{1-\alpha\rho}}\right)},
\end{aligned} \tag{42}$$

and by taking the inverse LT, we reach the solution as follows:

$$\begin{aligned}
\mathbf{q}(t) & = \mathbf{q}_0\frac{\delta}{\delta + CR\sigma^{1-\alpha\rho}}E_\alpha\left(-\frac{\sigma^{2(1-\alpha\rho)}}{\delta + CR\sigma^{1-\alpha\rho}}\left(\Gamma(\gamma+1)\frac{t^\rho}{\rho}\right)^\alpha\right) \\
& + \mathbf{q}_0\frac{CR\sigma^{1-\alpha\rho}}{\delta + CR\sigma^{1-\alpha\rho}}E_\alpha\left(-\frac{\sigma^{2(1-\alpha\rho)}}{\delta + CR\sigma^{1-\alpha\rho}}\left(\Gamma(\gamma+1)\frac{t^\rho}{\rho}\right)^\alpha\right) \\
& + \frac{C\sigma^{2(1-\alpha\rho)}\mathbf{e}_0}{\delta + CR\sigma^{1-\alpha\rho}}\left[1 - E_\alpha\left(-\frac{\sigma^{2(1-\alpha\rho)}}{\delta + CR\sigma^{1-\alpha\rho}}\left(\Gamma(\gamma+1)\frac{t^\rho}{\rho}\right)^\alpha\right)\right].
\end{aligned} \tag{43}$$

The plots for the solution Equation (43) is presented in Figures 13 and 14. We show the behavior of the function $\mathbf{q}(t)$ with constant source when $\alpha = 1, 0.9, 0.8, 0.7$, $\rho = 1, 0.9, 0.8, 0.7$, $\gamma = 1.2$ according to $R = 2\ \Omega$, $L = 10$ H, $C = 0.1$ F, $\mathbf{e}_0 = 5$ V, $\mathbf{q}(0) = 10$, and

$_M\mathcal{D}^{\alpha,\rho,\gamma}\mathbf{q}(0) = 0$. On the other hand, in Figure 14, one can see the solutions curves for some values of R and L when $\alpha = 0.7$, $\rho = 0.8$, and $\gamma = 0.9$. Moreover, Figures 13 and 14 show exponential behavior when R and L change for different values of fractional-orders. It is clear that fractional-orders have the power to increase or decrease wavelength and height as can be seen in Figures 15 and 16 showing oscillatory behavior under the damping system.

Case 2. (Exponential source). Let us assume that $\mathbf{e}(t) = \mathbf{e}_0 e^{-\lambda \Gamma(\gamma+1)\frac{t^\rho}{\rho}}$, $\mathbf{q}(0) = \mathbf{q}_0$, and $_M\mathcal{D}^{\rho,\gamma}\mathbf{q}(0) = 0$, we present Equation (37) as:

$$_M\mathcal{D}^{2\alpha,\rho,\gamma}\mathbf{q}(t) + \frac{CR}{\delta}\sigma^{1-\alpha\rho}{}_M\mathcal{D}^{\alpha,\rho,\gamma}\mathbf{q}(t) + \frac{\sigma^{2(1-\alpha\rho)}}{\delta}\mathbf{q}(t) = \frac{C\sigma^{2(1-\alpha\rho)}}{\delta}\mathbf{e}_0 e^{-\lambda\Gamma(\gamma+1)\frac{t^\rho}{\rho}}, \quad (44)$$

and taking the LT of the Equation (44), we can get:

$$\mathcal{L}_{\rho,\gamma}\{{}_M\mathcal{D}^{2\alpha,\rho,\gamma}\mathbf{q}(t)\} + \frac{CR\sigma^{1-\alpha\rho}}{\delta}\mathcal{L}_{\rho,\gamma}\{{}_M\mathcal{D}^{\alpha,\rho,\gamma}\mathbf{q}(t)\} + \frac{\sigma^{2(1-\alpha\rho)}}{\delta}\mathcal{L}_{\rho,\gamma}\{\mathbf{q}(t)\} = \frac{C\sigma^{2(1-\alpha\rho)}\mathbf{e}_0}{\delta}\mathcal{L}_{\rho,\gamma}\{e^{-\lambda\Gamma(\gamma+1)\frac{t^\rho}{\rho}}\}, \quad (45)$$

$$s^\alpha \mathcal{L}_{\rho,\gamma}\{\mathbf{q}(t)\} - s^{\alpha-1}\mathbf{q}(0) - s^{\alpha-2}{}_M\mathcal{D}^{\rho,\gamma}\mathbf{q}(0) \qquad (46)$$
$$+ \frac{CR\sigma^{1-\alpha\rho}}{\delta}\left[s^\alpha \mathcal{L}_{\rho,\gamma}\{\mathbf{q}(t)\} - s^{\alpha-1}\mathbf{q}_0\right] + \frac{\sigma^{2(1-\alpha\rho)}}{\delta}\mathcal{L}_{\rho,\gamma}\{\mathbf{q}(t)\}$$
$$= \frac{C\sigma^{2(1-\alpha\rho)}\mathbf{e}_0}{\delta(s+\lambda)},$$

$$\mathcal{L}_{\rho,\gamma}\{\mathbf{q}(t)\} = \mathbf{q}_0 \frac{\delta}{\delta + CR\sigma^{1-\alpha\rho}} \frac{s^{\alpha-1}}{s^\alpha + \frac{\sigma^{2(1-\alpha\rho)}}{\delta+CR\sigma^{1-\alpha\rho}}} + \mathbf{q}_0 \frac{CR\sigma^{1-\alpha\rho}}{\delta + CR\sigma^{1-\alpha\rho}} \frac{s^{\alpha-1}}{s^\alpha + \frac{\sigma^{2(1-\alpha\rho)}}{\delta+CR\sigma^{1-\alpha\rho}}} \qquad (47)$$
$$+ \frac{C\sigma^{2(1-\alpha\rho)}\mathbf{e}_0}{\delta + CR\sigma^{1-\alpha\rho}} \frac{1}{s+\lambda} \frac{1}{s^\alpha + \frac{\sigma^{2(1-\alpha\rho)}}{\delta+CR\sigma^{1-\alpha\rho}}},$$

$$\mathbf{q}(t) = \mathbf{q}_0 \frac{\delta}{\delta + CR\sigma^{1-\alpha\rho}} E_\alpha\left(-\frac{\sigma^{2(1-\alpha\rho)}}{\delta + CR\sigma^{1-\alpha\rho}}\left(\Gamma(\gamma+1)\frac{t^\rho}{\rho}\right)^\alpha\right)$$
$$+ \mathbf{q}_0 \frac{CR\sigma^{1-\alpha\rho}}{\delta + CR\sigma^{1-\alpha\rho}} E_\alpha\left(-\frac{\sigma^{2(1-\alpha\rho)}}{\delta + CR\sigma^{1-\alpha\rho}}\left(\Gamma(\gamma+1)\frac{t^\rho}{\rho}\right)^\alpha\right)$$
$$+ \frac{C\sigma^{2(1-\alpha\rho)}\mathbf{e}_0}{\delta + CR\sigma^{1-\alpha\rho}}\Gamma(\gamma+1)\int_0^t \left(\Gamma(\gamma+1)\frac{t^\rho}{\rho} - \Gamma(\gamma+1)\frac{\tau^\rho}{\rho}\right)^{\alpha-1} \qquad (48)$$
$$\times E_{\alpha,\alpha}\left(-\frac{\sigma^{2(1-\alpha\rho)}}{\delta + CR\sigma^{1-\alpha\rho}}\left(\Gamma(\gamma+1)\frac{t^\rho}{\rho} - \Gamma(\gamma+1)\frac{\tau^\rho}{\rho}\right)^\alpha\right)$$
$$\times \exp\left(-\lambda\Gamma(\gamma+1)\frac{\tau^\rho}{\rho}\right)\tau^{\rho-1}d\tau.$$

Figures 15 and 16 show the behavior of the solution Equation (48) under the exponential source for some values of α and ρ when $\gamma = 0.8$ and $\gamma = 1.5$, respectively. By deliberately choosing the α and ρ values the same on these two figures, we change the value of γ and clearly observe its effect on the system.

Case 3. (Oscillatory source). Supposing that $\mathbf{e}(t) = \mathbf{e}_0 \cos\left(\theta\Gamma(\gamma+1)\frac{t^\rho}{\rho}\right)$, $\mathbf{q}(0) = \mathbf{q}_0$, and $_M\mathcal{D}^{\rho,\gamma}\mathbf{q}(0) = 0$, we present Equation (37) as:

$$_M\mathcal{D}^{2\alpha,\rho,\gamma}\mathbf{q}(t) + \frac{CR}{\delta}\sigma^{1-\alpha\rho}{}_M\mathcal{D}^{\alpha,\rho,\gamma}\mathbf{q}(t) + \frac{\sigma^{2(1-\alpha\rho)}}{\delta}\mathbf{q}(t) = \frac{C\sigma^{2(1-\alpha\rho)}}{\delta}\mathbf{e}_0\cos\left(\theta\Gamma(\gamma+1)\frac{t^\rho}{\rho}\right), \quad (49)$$

by taking the LT of Equation (49), we attain the following relation:

$$\mathcal{L}_{\rho,\gamma}\{{}_M\mathcal{D}^{2\alpha,\rho,\gamma}\mathbf{q}(t)\} + \frac{CR\sigma^{1-\alpha\rho}}{\delta}\mathcal{L}_{\rho,\gamma}\{{}_M\mathcal{D}^{\alpha,\rho,\gamma}\mathbf{q}(t)\} + \frac{\sigma^{2(1-\alpha\rho)}}{\delta}\mathcal{L}_{\rho,\gamma}\{\mathbf{q}(t)\} \quad (50)$$
$$= \frac{C\sigma^{2(1-\alpha\rho)}\mathbf{e}_0}{\delta}\mathcal{L}_{\rho,\gamma}\left\{\cos\left(\theta\Gamma(\gamma+1)\frac{t^\rho}{\rho}\right)\right\},$$

$$s^\alpha \mathcal{L}_{\rho,\gamma}\{\mathbf{q}(t)\} - s^{\alpha-1}\mathbf{q}(0) - s^{\alpha-2}{}_M\mathbf{D}^{\rho,\gamma}\mathbf{q}(0)$$
$$+ \frac{CR\sigma^{1-\alpha\rho}}{\delta}\left[s^\alpha\mathcal{L}_{\rho,\gamma}\{\mathbf{q}(t)\} - s^{\alpha-1}\mathbf{q}_0\right] + \frac{\sigma^{2(1-\alpha\rho)}}{\delta}\mathcal{L}_{\rho,\gamma}\{\mathbf{q}(t)\} \quad (51)$$
$$= \frac{C\sigma^{2(1-\alpha\rho)}\mathbf{e}_0}{\delta}\frac{s}{\theta^2+s^2},$$

$$\mathcal{L}_{\rho,\gamma}\{\mathbf{q}(t)\} = \mathbf{q}_0 \frac{\delta}{\delta+CR\sigma^{1-\alpha\rho}}\frac{s^{\alpha-1}}{s^\alpha+\frac{\sigma^{2(1-\alpha\rho)}}{\delta+CR\sigma^{1-\alpha\rho}}} + \mathbf{q}_0\frac{CR\sigma^{1-\alpha\rho}}{\delta+CR\sigma^{1-\alpha\rho}}\frac{s^{\alpha-1}}{s^\alpha+\frac{\sigma^{2(1-\alpha\rho)}}{\delta+CR\sigma^{1-\alpha\rho}}} \quad (52)$$
$$+ \frac{C\sigma^{2(1-\alpha\rho)}\mathbf{e}_0}{\delta+CR\sigma^{1-\alpha\rho}}\frac{s}{\theta^2+s^2}\frac{1}{s^\alpha+\frac{\sigma^{2(1-\alpha\rho)}}{\delta+CR\sigma^{1-\alpha\rho}}},$$

$$\mathbf{q}(t) = \mathbf{q}_0 \frac{\delta}{\delta+CR\sigma^{1-\alpha\rho}}E_\alpha\left(-\frac{\sigma^{2(1-\alpha\rho)}}{\delta+CR\sigma^{1-\alpha\rho}}\left(\Gamma(\gamma+1)\frac{t^\rho}{\rho}\right)^\alpha\right)$$
$$+ \mathbf{q}_0 \frac{CR\sigma^{1-\alpha\rho}}{\delta+CR\sigma^{1-\alpha\rho}}E_\alpha\left(-\frac{\sigma^{2(1-\alpha\rho)}}{\delta+CR\sigma^{1-\alpha\rho}}\left(\Gamma(\gamma+1)\frac{t^\rho}{\rho}\right)^\alpha\right)$$
$$+ \frac{C\sigma^{2(1-\alpha\rho)}\mathbf{e}_0}{\delta+CR\sigma^{1-\alpha\rho}}\Gamma(\gamma+1)\int_0^t\left(\Gamma(\gamma+1)\frac{t^\rho}{\rho}-\Gamma(\gamma+1)\frac{\tau^\rho}{\rho}\right)^{\alpha-1} \quad (53)$$
$$\times E_{\alpha,\alpha}\left(-\frac{\sigma^{2(1-\alpha\rho)}}{\delta+CR\sigma^{1-\alpha\rho}}\left(\Gamma(\gamma+1)\frac{t^\rho}{\rho}-\Gamma(\gamma+1)\frac{\tau^\rho}{\rho}\right)^\alpha\right)$$
$$\times \cos\left(\theta\Gamma(\gamma+1)\frac{t^\rho}{\rho}\right)\tau^{\rho-1}d\tau.$$

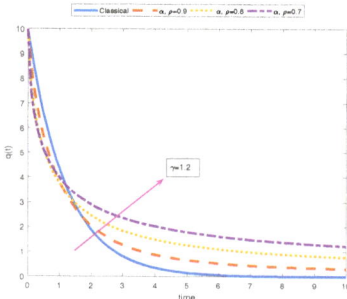

Figure 13. This plot is for the function $\mathbf{q}(t)$ with respect to the constant source for some values of α and ρ when $\gamma = 1.2$, $R = 2\,\Omega$, $L = 10$ H, $C = 0.1$ F, $\mathbf{e}_0 = 5$ V, $\mathbf{q}(0) = 10$, and ${}_M\mathcal{D}^{\alpha,\rho,\gamma}\mathbf{q}(0) = 0$.

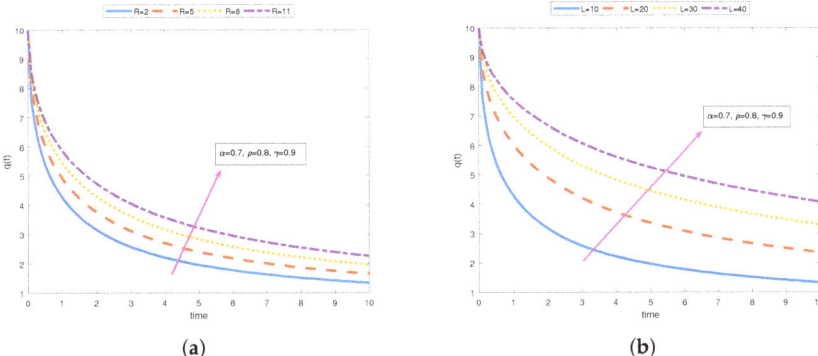

Figure 14. The figure (**a**) is for the function $\mathbf{q}(t)$ with respect to the constant source for some values of α, ρ, and γ in order to see the impact of R, and the figure (**b**) is plotted to see the effect of L when $R = 2\,\Omega$, $L = 10$ H, $C = 0.1$ F, $\mathbf{e}_0 = 5$ V, $\mathbf{q}(0) = 10$, and $_M\mathcal{D}^{\alpha,\rho,\gamma}\mathbf{q}(0) = 0$.

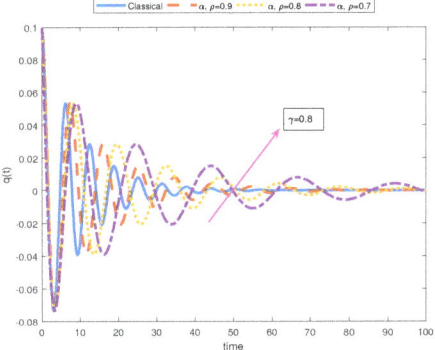

Figure 15. This plot is for the function $\mathbf{q}(t)$ with respect to the exponential source for some values of α and ρ when $\gamma = 0.8$, $\mathbf{e}_0 = 5$ V, $L = 10$ H, $C = 0.1$ F, $R = 2\,\Omega$, $\mathbf{q}(0) = 10$, and $_M\mathcal{D}^{\alpha,\rho,\gamma}\mathbf{q}(0) = 0$.

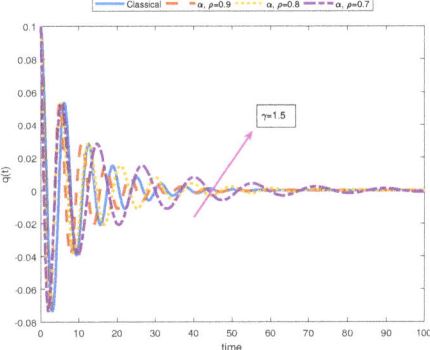

Figure 16. This plot is for the function $\mathbf{q}(t)$ with respect to the exponential source for some values of α and ρ when $\gamma = 1.5$, $\mathbf{e}_0 = 5$ V, $L = 10$ H, $C = 0.1$ F, $R = 2\,\Omega$, $\mathbf{q}(0) = 10$, and $_M\mathcal{D}^{\alpha,\rho,\gamma}\mathbf{q}(0) = 0$.

3. Comparative Analysis and Concluding Remarks

Here, we have performed a comparative analysis to observe the impact of the additional parameters inside the non-local fractional M-derivative. For this purpose, we have compared our results with the solution obtained via the Caputo operator in [8]. This comparison has been carried out for the RC circuit with constant source by employing the experimental data obtained from the electronic laboratory in CENIDET. The used experimental data is $R = 10\ \Omega$, $C = 1000$ F, $e_0 = 7.58$, and $\mathbf{V}_c(0) = 0$ and for the second case, $R = 10\ \Omega$, $C = 1000$ F, $e_0 = 0$, and $\mathbf{V}_c(0) = 7.58$ as seen in [8]. The values for Figures 17 and 18 are as follows: $R = 10\ \Omega$, $C = 1000$ F, $e_0 = 7.58$, $\mathbf{V}_c(0) = 0$, and $R = 10\ \Omega$, $C = 1000$ F, $e_0 = 0$, and $\mathbf{V}_c(0) = 7.58$ for Figures 19 and 20. We observe that the non-local fractional M-derivative behaves closer to the experimental data than the Caputo derivative thanks to the convenient values of ρ and γ. Figures 17 and 19 have been plotted for $\alpha = 0.9$, $\rho = 1.8$, and $\gamma = 1.8$ while Figures 18 and 20 have shown when $\alpha = 0.7$, $\rho = 1.5$, and $\gamma = 2$. It should be noted that the non-local M-derivative perform the same behavior with the Caputo fractional derivative when $\alpha = 1$, $\rho = 1$, and $\gamma = 1$. On the other hand, the Caputo derivative tends faster to the steady-state than the non-local M-derivative and traditional counterpart in Figures 17–20 having exponentially dynamics.

Moreover, some general conclusions on our main results can be listed as below:

- We have carried out an efficient extension of a physical problem through a non-local singular fractional operator by providing the solutions including three arbitrary parameters α, ρ, and γ;
- A detailed analysis has been introduced for the Resistance-Capacitance (RC), Inductance-Capacitance (LC), and Resistance-Inductance-Capacitance (RLC) electric circuits utilizing a generalized type fractional operator in the sense of Caputo called non-local M-derivative;
- Due to the fact that all solutions obtained in this study depend on three parameters unlike the other studies in the literature, the solutions we have obtained are more general results;
- In order to show the benefits of the non-local M-derivative for the proposed physical problem, a comprehensive comparison has been addressed for the RC circuit with constant source in the light of experimental data;
- As a result of our observations on Figures 1–16, we see that the amplitudes get smaller or grow for some increasing or decreasing values of α, ρ, and γ. The waves also displace as α, ρ, and γ change;
- Importantly, the arbitrary parameters α, ρ, and γ allow us to get some crucial information about the intrinsic properties of the problem under investigation.

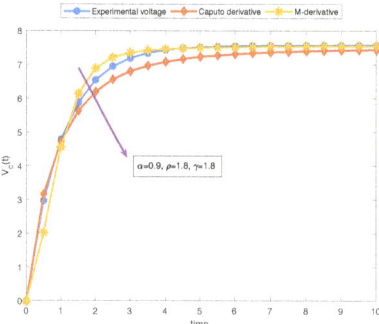

Figure 17. Comparison of the non-local M-derivative and Caputo derivative with experimental data for different values of α, ρ, and γ under the Resistance-Capacitance (RC) circuit including constant source when $\alpha = 0.9$, $\rho = 1.8$, $\gamma = 1.8$, $R = 10\ \Omega$, $C = 1000$ F, $e_0 = 7.58$, and $\mathbf{V}_c(0) = 0$.

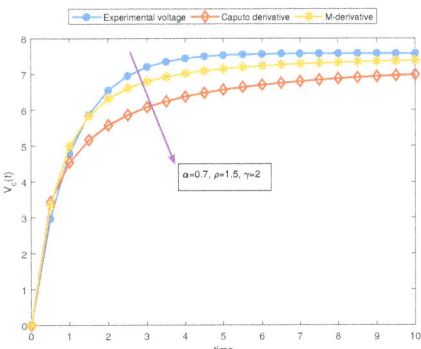

Figure 18. Comparison of the non-local M-derivative and Caputo derivative with experimental data for different values of α, ρ, and γ under the RC circuit including constant source when $\alpha = 0.7$, $\rho = 1.5$, $\gamma = 2$, $R = 10\ \Omega$, $C = 1000$ F, $\mathbf{e}_0 = 7.58$, and $\mathbf{V}_c(0) = 0$.

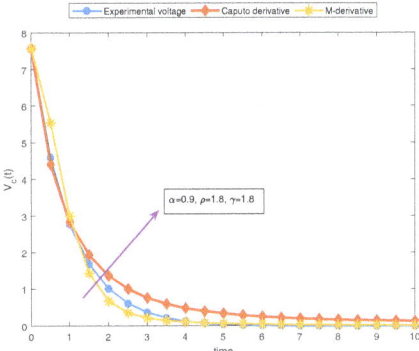

Figure 19. Comparison of the non-local M-derivative and Caputo derivative with experimental data for different values of α, ρ, and γ under the RC circuit including constant source when $\alpha = 0.9$, $\rho = 1.8$, $\gamma = 1.8$, $R = 10\ \Omega$, $C = 1000$ F, $\mathbf{e}_0 = 0$, and $\mathbf{V}_c(0) = 7.58$.

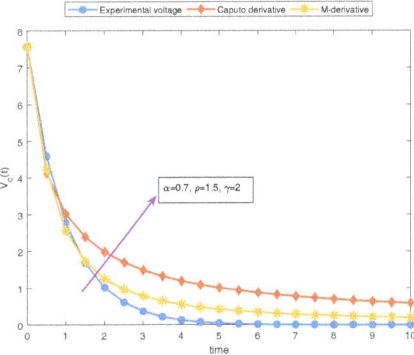

Figure 20. Comparison of the non-local M-derivative and Caputo derivative with experimental data for different values of α, ρ, and γ under the RC circuit including constant source when $\alpha = 0.7$, $\rho = 1.5$, $\gamma = 2$, $R = 10\ \Omega$, $C = 1000$ F, $\mathbf{e}_0 = 0$, and $\mathbf{V}_c(0) = 7.58$.

Author Contributions: Validation, M.I.; writing—review and editing, B.A. All authors have read and agreed to the published version of the manuscript.

Funding: This research received no external funding.

Data Availability Statement: Not applicable.

Conflicts of Interest: The authors declare no conflict of interest.

References

1. Kilbas, A.A.; Srivastava, H.M.; Trujillo, J.J. *Theory and Applications of Fractional Differential Equations*; Elsevier: Amsterdam, The Netherlands, 2006; p. 204.
2. Katugampola, U.N. New approach to a generalized fractional integral. *Appl. Math. Comput.* **2011**, *218*, 860–865. [CrossRef]
3. Acay, B.; Bas, E.; Abdeljawad, T. Non-local fractional calculus from different viewpoint generated by truncated M-derivative. *J. Comput. Appl. Math.* **2020**, *366*, 112410. [CrossRef]
4. Sousa, J.V.D.C.; de Oliveira, E.C. A new truncated M-fractional derivative type unifying some fractional derivative types with classical properties. *arXiv* **2017**, arXiv:1704.08187.
5. Jarad, F.; Abdeljawad, T.; Alzabut, J. Generalized fractional derivatives generated by a class of local proportional derivatives. *Eur. Phys. J. Spec. Top.* **2017**, *226*, 3457–3471. [CrossRef]
6. Baleanu, D.; Fernandez, A.; Akgül, A. On a Fractional Operator Combining Proportional and Classical Differintegrals. *Mathematics* **2020**, *8*, 360. [CrossRef]
7. Gómez-Aguilar, J.F.; Yépez-Martínez, H.; Escobar-Jiménez, R.F.; Astorga-Zaragoza, C.M.; Reyes-Reyes, J. Analytical and numerical solutions of electrical circuits described by fractional derivatives. *Appl. Math. Model.* **2016**, *40*, 9079–9094. [CrossRef]
8. Gómez-Aguilar, J.F. Fundamental solutions to electrical circuits of non-integer order via fractional derivatives with and without singular kernels. *Eur. Phys. J. Plus* **2018**, *133*, 197. [CrossRef]
9. Gómez-Aguilar, J.F.; Atangana, A.; Morales-Delgado, V.F. Electrical circuits RC, LC, and RL described by Atangana–Baleanu fractional derivatives. *Int. J. Circuit Theory Appl.* **2017**, *45*, 1514–1533. [CrossRef]
10. Abro, K.A.; Memon, A.A.; Uqaili, M.A. A comparative mathematical analysis of RL and RC electrical circuits via Atangana-Baleanu and Caputo-Fabrizio fractional derivatives. *Eur. Phys. J. Plus* **2018**, *133*, 1–9. [CrossRef]
11. Sene, N.; Gómez-Aguilar, J.F. Analytical solutions of electrical circuits considering certain generalized fractional derivatives. *Eur. Phys. J. Plus* **2019**, *134*, 1–14. [CrossRef]
12. Martinez, L.; Rosales, J.J.; Carreño, C.A.; Lozano, J.M. Electrical circuits described by fractional conformable derivative. *Int. J. Circuit Theory Appl.* **2018**, *46*, 1091–1100. [CrossRef]
13. Jarad, F.; Ugurlu, E.; Abdeljawad, T.; Baleanu, D. On a new class of fractional operators. *Adv. Differ. Equ.* **2017**, *2017*, 247. [CrossRef]
14. Baleanu, D.; Fernandez, A. On fractional operators and their classifications. *Mathematics* **2019**, *7*, 830. [CrossRef]
15. Qureshi, S.; Yusuf, A.; Shaikh, A.A.; Inc, M.; Baleanu, D. Fractional modeling of blood ethanol concentration system with real data application. *Chaos Interdiscip. J. Nonlinear Sci.* **2019**, *29*, 013143. [CrossRef] [PubMed]
16. Qureshi, S.; Yusuf, A. Modeling chickenpox disease with fractional derivatives: From caputo to atangana-baleanu. *Chaos Solitons Fractals* **2019**, *122*, 111–118. [CrossRef]
17. Acay, B.; Bas, E.; Abdeljawad, T. Fractional economic models based on market equilibrium in the frame of different type kernels. *Chaos Solitons Fractals* **2020**, *130*, 109438. [CrossRef]
18. Bas, E.; Acay, B.; Ozarslan, R. The price adjustment equation with different types of conformable derivatives in market equilibrium. *AIMS Math.* **2019**, *4*, 805–820. [CrossRef]
19. Acay, B.; Ozarslan, R.; Bas, E. Fractional physical models based on falling body problem. *AIMS Math.* **2020**, *5*, 2608. [CrossRef]
20. Yavuz, M. Novel solution methods for initial boundary value problems of fractional order with conformable differentiation. *Int. J. Optim. Control Theor. Appl. IJOCTA* **2017**, *8*, 1–7. [CrossRef]
21. Yusuf, A.; Inc, M.; Aliyu, A.I. On dark optical solitons of the space time nonlinear Schrödinger equation with fractional complex transform for Kerr and power law nonlinearities. *J. Coupled Syst. Multiscale Dyn.* **2018**, *6*, 114–120. [CrossRef]
22. Ozarslan, R. Microbial survival and growth modeling in frame of nonsingular fractional derivatives. *Math. Methods Appl. Sci.* **2020**, *2020*, 1–19. [CrossRef]
23. Yavuz, M.; Abdeljawad, T. Nonlinear regularized long-wave models with a new integral transformation applied to the fractional derivative with power and Mittag-Leffler kernel. *Adv. Differ. Equ.* **2020**, *2020*, 367. [CrossRef]
24. Yavuz, M.; Sene, N. Stability Analysis and Numerical Computation of the Fractional Predator–Prey Model with the Harvesting Rate. *Fractal Fract.* **2020**, *4*, 35. [CrossRef]
25. Jarad, F.; Abdeljawad, T. Generalized fractional derivatives and Laplace transform. *Discret. Contin. Dyn. Syst. S* **2020**, *13*, 709. [CrossRef]
26. Ozarslan, R.; Bas, E.; Baleanu, D.; Acay, B. Fractional physical problems including wind-influenced projectile motion with Mittag-Leffler kernel. *AIMS Math.* **2020**, *5*, 467. [CrossRef]

 *fractal and fractional*

Article

A Fractional SAIDR Model in the Frame of Atangana–Baleanu Derivative

Esmehan Uçar, Sümeyra Uçar, Fırat Evirgen and Necati Özdemir *

Department of Mathematics, Faculty of Arts and Sciences, Balıkesir University, Balıkesir 10145, Turkey; esucarr@gmail.com (E.U.); sumeyraucar@balikesir.edu.tr (S.U.); fevirgen@balikesir.edu.tr (F.E.)
* Correspondence: nozdemir@balikesir.edu.tr

Abstract: It is possible to produce mobile phone worms, which are computer viruses with the ability to command the running of cell phones by taking advantage of their flaws, to be transmitted from one device to the other with increasing numbers. In our day, one of the services to gain currency for circulating these malignant worms is SMS. The distinctions of computers from mobile devices render the existing propagation models of computer worms unable to start operating instantaneously in the mobile network, and this is particularly valid for the SMS framework. The susceptible–affected–infectious–suspended–recovered model with a classical derivative (abbreviated as SAIDR) was coined by Xiao et al., (2017) in order to correctly estimate the spread of worms by means of SMS. This study is the first to implement an Atangana–Baleanu (AB) derivative in association with the fractional SAIDR model, depending upon the SAIDR model. The existence and uniqueness of the drinking model solutions together with the stability analysis are shown through the Banach fixed point theorem. The special solution of the model is investigated using the Laplace transformation and then we present a set of numeric graphics by varying the fractional-order θ with the intention of showing the effectiveness of the fractional derivative.

Keywords: fractional differential equations; fixed point theory; Atangana–Baleanu derivative; mobile phone worms

MSC: 34A08; 47H10; 34A34

1. Introduction

Although computer worms are collected under the category of computer viruses, they can be treated as a separate group owing to their distinct characteristics. First and foremost, our intervention is not needed for computer worms to transmit whereas it is a must for viruses, as they need a user to have access to an electronic document, directive, or software, etc. In addition, computer worms are able to autonomously transmit themselves and are also capable of producing replicas of themselves; this grants worms the capacity to generate numerous duplicates for being transmitted to and infecting other computers.

Mobile worms have become increasingly contagious in parallel with the immense expansion of the cellular network system and the growing demand on mobile phones. The majority of these worms carry the potential to cause irrepairable damages to the mobile domain; for example, it is quite likely that private information can be seized, collected, or leaked from an infected device by computer worms. Furthermore, the fact that the smart phones available in the market today are open to plenty of security breaches entails probable widespread infections by the mobile malware in question, which carries a significant risk. In the meantime, people employ many diverse means to circulate various electronic documents, participate in a variety of pursuits, or attend gatherings on the Internet with the smart phones at their disposal, and these practices call forth the invasion of mobile devices by worms. Therefore, SMS has also become one of the typical system components via which worms are transmitted. A term called an "SMS-based worm", which is a variety of mobile worms, has been coined in the literature following the example of Sea [1], Cckun [2], Selfmite [3], and xxShenQi [4], which are the most prominent examples of relevance amongst others.

In order to assess the impact of memory on computer models, fractional calculus rises to prominence, which also yields more accurate outcomes. That is to say, fractional calculus is more versatile compared to classical calculus owing to hereditary features and the definition of memory. Caputo [5], Liouville-Caputo [6], as well as Caputo and Fabrizio [7] set forth a great deal of conceptions concerning fractional order operators and these conceptions have been proven quite efficacious when devising representations of many real-world problems [8–19]. In addition, the said derivatives have been proven quite efficient when one adopts numerical methods and examines the relation between distinct problems by means of comparison. It can be seen in a number of studies that employing fractional order derivatives yields more successful outcomes in terms of acquiring real data for distinct worm models [20,21]. Performed according to the principal of a generalised Mittag–Leffler function in the role of a non-singular and non-local kernel, an innovative fractional order derivative was brought into operation for the first time by Atangana and Baleanu [22] in 2016. This newly defined Atangana–Baleanu (AB) derivative obtains better results in many actual problems [22–28].

The purpose of this work was to delve into a susceptible–affected–infectious–suspended–recovered (SAIDR) model, a type of fractional order model which, for the first time, was put forward in [29] employing a classical derivative aimed at SMS-based worm propagation in mobile networks, on the basis of more favourable fractional calculus theories. As long as susceptible users of mobile devices refrain from opening the links that are harmful, it is not possible for them to instantly enter into the infected state even if the malicious message is delivered to the said devices. This is the reason behind the addition of the affected state into [29] by its authors; abbreviated as state $A(t)$, it delineates the circumstance when a harmful link is delivered to a user but not yet opened. What is more, particularly if the phone is damaged, the harmful message is not invariably circulated by an infected node. Thus, another new state is also instituted, which is the suspended state, going by the abbreviation of state $D(t)$. This is of a unique quality since a harmful message cannot be circulated by an infected smart phone in spite of its existence in the given device. Lastly, the overall quantity of worm nodes are separated as follows: $N = S(t) + A(t) + I(t) + D(t) + R(t)$, i.e., $S(t)$ susceptible state, $A(t)$ affected state, $I(t)$ infected state, $D(t)$ suspended state, $R(t)$ recovered state. The integer order differential equation system which puts forth the SAIDR model in [29] can be seen below:

$$\begin{aligned}
\frac{dS(t)}{dt} &= \mu N - \gamma S - \beta S I - \mu S, \\
\frac{dA(t)}{dt} &= \beta S I - \delta A - \eta A - \mu A, \\
\frac{dI(t)}{dt} &= \eta A - \sigma I - \tau I - \mu I, \\
\frac{dD(t)}{dt} &= \tau I - \varphi D - \mu D, \\
\frac{dR(t)}{dt} &= \gamma S + \delta A + \sigma I + \varphi D - \mu R.
\end{aligned} \quad (1)$$

The variable factors concerning the model alter at time t as follows: the susceptible node is converted into the affected state by β, the ratio of infection, when a harmful electronic message is delivered to it from another point of intersection. The worm is transmitted to the node which is in the affected state by a ratio of η in the event that it is rendered active by the affected node opening the malignant link enclosed in the message. The node shifts into the suspended state from the infected state by a transition ratio of τ. Meanwhile, certain software against malware might be set up in mobile devices in order to block or erase harmful messages. Once the said software are set up, the phone cannot permanently remain in the infected state. These safety software can also be set up in the wake of a maintenance process, provided that the phone stays in the special suspended state. Hence, the node is able to shift into the ultimate recovered state regardless of its current state. The ratios by which the node recovers from states $S(t)$, $A(t)$, $I(t)$, and $D(t)$ to state $R(t)$ are denominated as γ, δ, σ, φ in the order given.

The article is further structured with the subdivisions specified below: in order to clarify the remaining body of this work, several fundamental notions are introduced in Section 2. Section 3 substantiates the existence and uniqueness of the solution for the proposed model, while we scrutinize the specific solution for the model along with the Laplace transformation and approach the stability analysis concerning the technique by means of the fixed point principle in Section 4. In Section 5, this fractional order model is numerically depicted so as to review the total effect. Finally, we bring our study to an end by debating the acquired outcomes.

2. Some Preliminaries

Here, we recall some fundamental notions.

Definition 1. *The ABR fractional derivative (R denotes Riemann–Liouville type) is defined by [30]*

$$^{ABR}D^{\theta}_{a^+}[f(t)] = \frac{F(\theta)}{1-\theta} \frac{d}{dt} \int_a^t f'(x) E_{\theta}\left[\frac{-\theta}{1-\theta}(t-x)^{\theta}\right] dx \quad (2)$$

for $0 < \theta < 1$, $a < t < b$ and $f \in L^1(a,b)$.

Definition 2. *The ABC fractional derivative (C denotes Caputo type) is defined by [30]*

$$^{ABC}D^{\theta}_{a^+}[f(t)] = \frac{F(\theta)}{1-\theta} \int_a^t f'(x) E_{\theta}\left[\frac{-\theta}{1-\theta}(t-x)^{\theta}\right] dx \quad (3)$$

for $0 < \theta < 1$, $a < t < b$ and f a differentiable function on $[a,b]$ such that $f' \in L^1(a,b)$.

Definition 3. *The AB fractional integral operator $^{AB}I^{\theta}_{a^+}$ defined by [30]*

$$^{AB}I^{\theta}_{a^+}f(t) = \frac{1-\theta}{F(\theta)}f(t) + \frac{\theta}{F(\theta)}{^{RL}I^{\theta}_{a^+}}f(t). \quad (4)$$

In the above definitions, the function E_{θ} is the Mittag–Leffler function given by

$$E_{\theta} = \sum_{n=0}^{\infty} \frac{x^n}{\Gamma(\theta n + 1)}. \quad (5)$$

3. Existence of a Unique Solution

In the present work, we enlarged the model (1) by substituting the time derivative by the Atangana–Baleanu derivative. With this change, the right- and left-hand sides will not have the same dimensions. To overcome this matter, we used an auxiliary parameter κ with the dimension of s, to change the fractional operator so that the sides have the same dimension [31]. Thereby, we give the following fractional system:

$$\frac{1}{\kappa^{1-\theta}} {}_{0}^{ABC}D_{t}^{\theta}S(t) = \mu N - \gamma S - \beta SI - \mu S,$$

$$\frac{1}{\kappa^{1-\theta}} {}_{0}^{ABC}D_{t}^{\theta}A(t) = \beta SI - \delta A - \eta A - \mu A,$$

$$\frac{1}{\kappa^{1-\theta}} {}_{0}^{ABC}D_{t}^{\theta}I(t) = \eta A - \sigma I - \tau I - \mu I, \qquad (6)$$

$$\frac{1}{\kappa^{1-\theta}} {}_{0}^{ABC}D_{t}^{\theta}D(t) = \tau I - \varphi D - \mu D,$$

$$\frac{1}{\kappa^{1-\theta}} {}_{0}^{ABC}D_{t}^{\theta}R(t) = \gamma S + \delta A + \sigma I + \varphi D - \mu R.$$

with the initial numbers $S(0) = S_0, A(0) = A_0, I(0) = I_0, D(0) = D_0, R(0) = R_0$ where ${}_{a}^{ABC}D_{t}^{\theta}$ is the AB derivative in Caputo type and $\theta \in [0,1]$.

In this part, we prove that the system (6) has a unique solution. Implementing the fractional integral into the system (6) by handling the Corollary 2.3 in [30], we have:

$$S(t) - S(0) = \frac{(1-\theta)\kappa^{1-\theta}}{F(\theta)}[\mu N - \gamma S(t) - \beta S(t)I(t) - \mu S(t)]$$

$$+ \frac{\theta \kappa^{1-\theta}}{F(\theta)\Gamma(\theta)} \int_0^t (t-\lambda)^{\theta-1}[\mu N - \gamma S(\lambda) - \beta S(\lambda)I(\lambda) - \mu S(\lambda)]d\lambda,$$

$$A(t) - A(0) = \frac{(1-\theta)\kappa^{1-\theta}}{F(\theta)}[\beta S(t)I(t) - \delta A(t) - \eta A(t) - \mu A(t)]$$

$$+ \frac{\theta \kappa^{1-\theta}}{F(\theta)\Gamma(\theta)} \int_0^t (t-\lambda)^{\theta-1}[\beta S(\lambda)I(\lambda) - \delta A(\lambda) - \eta A(\lambda) - \mu A(\lambda)]d\lambda,$$

$$I(t) - I(0) = \frac{(1-\theta)\kappa^{1-\theta}}{F(\theta)}[\eta A(t) - \sigma I(t) - \tau I(t) - \mu I(t)] \qquad (7)$$

$$+ \frac{\theta \kappa^{1-\theta}}{F(\theta)\Gamma(\theta)} \int_0^t (t-\lambda)^{\theta-1}[\eta A(\lambda) - \sigma I(\lambda) - \tau I(\lambda) - \mu I(\lambda)]d\lambda,$$

$$D(t) - D(0) = \frac{(1-\theta)\kappa^{1-\theta}}{F(\theta)}[\tau I(t) - \varphi D(t) - \mu D(t)]$$

$$+ \frac{\theta \kappa^{1-\theta}}{F(\theta)\Gamma(\theta)} \int_0^t (t-\lambda)^{\theta-1}[\tau I(\lambda) - \varphi D(\lambda) - \mu D(\lambda)]d\lambda,$$

$$R(t) - R(0) = \frac{(1-\theta)\kappa^{1-\theta}}{F(\theta)}[\gamma S(t) + \delta A(t) + \sigma I(t) + \varphi D(t) - \mu R(t)]$$

$$+ \frac{\theta \kappa^{1-\theta}}{F(\theta)\Gamma(\theta)} \int_0^t (t-\lambda)^{\theta-1}[\gamma S(\lambda) + \delta A(\lambda) + \sigma I(\lambda) + \varphi D(\lambda) - \mu R(\lambda)]d\lambda.$$

Let:
$$P_1(t,S) = \mu N - \gamma S(t) - \beta S(t)I(t) - \mu S(t),$$
$$P_2(t,A) = \beta S(t)I(t) - \delta A(t) - \eta A(t) - \mu A(t),$$
$$P_3(t,I) = \eta A(t) - \sigma I(t) - \tau I(t) - \mu I(t), \qquad (8)$$
$$P_4(t,D) = \tau I(t) - \varphi D(t) - \mu D(t),$$
$$P_5(t,R) = \gamma S(t) + \delta A(t) + \sigma I(t) + \varphi D(t) - \mu R(t).$$

Theorem 1. *The kernel P_1 satisfies the Lipschitz condition and contraction if the following inequality holds:*
$$0 < \gamma + \mu + \beta c \leq 1.$$

Proof of Theorem 1. Let S and S_1 be two functions, we have:
$$\|P_1(t,S) - P_1(t,S_1)\| = \|\gamma(S(t) - S_1(t)) + \beta I(t)(S(t) - S_1(t)) + \mu(S(t) - S_1(t))\|$$
$$\leq [\gamma + \mu + \beta\|I(t)\|]\|S(t) - S_1(t)\|.$$

Taking $\varepsilon_1 = \gamma + \mu + \beta c$ where $\|S(t)\| \leq a, \|A(t)\| \leq b, \|I(t)\| \leq c, \|D(t)\| \leq d, \|R(t)\| \leq e$ are bounded functions. Then, we find:
$$\|P_1(t,S) - P_1(t,S_1)\| \leq \varepsilon_1 \|S(t) - S_1(t)\|. \tag{9}$$

Hence, we find that the Lipschitz condition is provided by P_1 and since $0 < \gamma + \mu + \beta c \leq 1$, P_1 is also a contraction. □

Similarly, the other kernels P_2, P_3, P_4 and P_5 satisfy the Lipschitz condition and contraction:
$$\begin{aligned}
\|P_2(t,A) - P_2(t,A_1)\| &\leq \varepsilon_2 \|A(t) - A_1(t)\|, \\
\|P_3(t,I) - P_3(t,I_1)\| &\leq \varepsilon_3 \|I(t) - I_1(t)\|, \\
\|P_4(t,D) - P_4(t,D_1)\| &\leq \varepsilon_4 \|D(t) - D_1(t)\|, \\
\|P_5(t,R) - P_5(t,R_1)\| &\leq \varepsilon_5 \|R(t) - R_1(t)\|.
\end{aligned} \tag{10}$$

Regarding kernels P_1, P_2, P_3, P_4, P_5, Equation (7) becomes:

$$S(t) = S(0) + \frac{(1-\theta)\kappa^{1-\theta}}{F(\theta)} P_1(t,S) + \frac{\theta \kappa^{1-\theta}}{F(\theta)\Gamma(\theta)} \int_0^t (t-\lambda)^{\theta-1} P_1(\lambda,S) d\lambda,$$

$$A(t) = A(0) + \frac{(1-\theta)\kappa^{1-\theta}}{F(\theta)} P_2(t,A) + \frac{\theta \kappa^{1-\theta}}{F(\theta)\Gamma(\theta)} \int_0^t (t-\lambda)^{\theta-1} P_2(\lambda,A) d\lambda,$$

$$I(t) = I(0) + \frac{(1-\theta)\kappa^{1-\theta}}{F(\theta)} P_3(t,I) + \frac{\theta \kappa^{1-\theta}}{F(\theta)\Gamma(\theta)} \int_0^t (t-\lambda)^{\theta-1} P_3(\lambda,I) d\lambda, \tag{11}$$

$$D(t) = D(0) + \frac{(1-\theta)\kappa^{1-\theta}}{F(\theta)} P_4(t,D) + \frac{\theta \kappa^{1-\theta}}{F(\theta)\Gamma(\theta)} \int_0^t (t-\lambda)^{\theta-1} P_4(\lambda,D) d\lambda,$$

$$R(t) = R(0) + \frac{(1-\theta)\kappa^{1-\theta}}{F(\theta)} P_5(t,R) + \frac{\theta \kappa^{1-\theta}}{F(\theta)\Gamma(\theta)} \int_0^t (t-\lambda)^{\theta-1} P_5(\lambda,R) d\lambda.$$

Considering Equation (11) for the following recursive formula:

$$S_n(t) = \frac{(1-\theta)\kappa^{1-\theta}}{F(\theta)} P_1(t, S_{n-1}) + \frac{\theta \kappa^{1-\theta}}{F(\theta)\Gamma(\theta)} \int_0^t (t-\lambda)^{\theta-1} P_1(\lambda, S_{n-1}) d\lambda,$$

$$A_n(t) = \frac{(1-\theta)\kappa^{1-\theta}}{F(\theta)} P_2(t, A_{n-1}) + \frac{\theta \kappa^{1-\theta}}{F(\theta)\Gamma(\theta)} \int_0^t (t-\lambda)^{\theta-1} P_2(\lambda, A_{n-1}) d\lambda,$$

$$I_n(t) = \frac{(1-\theta)\kappa^{1-\theta}}{F(\theta)} P_3(t, I_{n-1}) + \frac{\theta \kappa^{1-\theta}}{F(\theta)\Gamma(\theta)} \int_0^t (t-\lambda)^{\theta-1} P_3(\lambda, I_{n-1}) d\lambda, \quad (12)$$

$$D_n(t) = \frac{(1-\theta)\kappa^{1-\theta}}{F(\theta)} P_4(t, D_{n-1}) + \frac{\theta \kappa^{1-\theta}}{F(\theta)\Gamma(\theta)} \int_0^t (t-\lambda)^{\theta-1} P_4(\lambda, D_{n-1}) d\lambda,$$

$$R_n(t) = \frac{(1-\theta)\kappa^{1-\theta}}{F(\theta)} P_5(t, R_{n-1}) + \frac{\theta \kappa^{1-\theta}}{F(\theta)\Gamma(\theta)} \int_0^t (t-\lambda)^{\theta-1} P_5(\lambda, R_{n-1}) d\lambda.$$

where $S_0(t) = S(0), A_0(t) = A(0), I_0(t) = I(0), D_0(t) = D(0), R_0(t) = R(0)$.
We deal with the difference between successive terms as below:

$$\Phi_{1n}(t) = S_n(t) - S_{n-1}(t) = \frac{(1-\theta)\kappa^{1-\theta}}{F(\theta)} [P_1(t, S_{n-1}) - P_1(t, S_{n-2})]$$
$$+ \frac{\theta \kappa^{1-\theta}}{F(\theta)\Gamma(\theta)} \int_0^t (t-\lambda)^{\theta-1} [P_1(\lambda, S_{n-1}) - P_1(\lambda, S_{n-2})] d\lambda,$$

$$\Phi_{2n}(t) = A_n(t) - A_{n-1}(t) = \frac{(1-\theta)\kappa^{1-\theta}}{F(\theta)} [P_2(t, A_{n-1}) - P_2(t, A_{n-2})]$$
$$+ \frac{\theta \kappa^{1-\theta}}{F(\theta)\Gamma(\theta)} \int_0^t (t-\lambda)^{\theta-1} [P_2(\lambda, A_{n-1}) - P_2(\lambda, A_{n-2})] d\lambda,$$

$$\Phi_{3n}(t) = I_n(t) - I_{n-1}(t) = \frac{(1-\theta)\kappa^{1-\theta}}{F(\theta)} [P_3(t, I_{n-1}) - P_3(t, I_{n-2})] \quad (13)$$
$$+ \frac{\theta \kappa^{1-\theta}}{F(\theta)\Gamma(\theta)} \int_0^t (t-\lambda)^{\theta-1} [P_3(\lambda, I_{n-1}) - P_3(\lambda, I_{n-2})] d\lambda,$$

$$\Phi_{4n}(t) = D_n(t) - D_{n-1}(t) = \frac{(1-\theta)\kappa^{1-\theta}}{F(\theta)} [P_4(t, D_{n-1}) - P_4(t, D_{n-2})]$$
$$+ \frac{\theta \kappa^{1-\theta}}{F(\theta)\Gamma(\theta)} \int_0^t (t-\lambda)^{\theta-1} [P_4(\lambda, D_{n-1}) - P_4(\lambda, D_{n-2})] d\lambda,$$

$$\Phi_{5n}(t) = R_n(t) - R_{n-1}(t) = \frac{(1-\theta)\kappa^{1-\theta}}{F(\theta)} [P_5(t, R_{n-1}) - P_5(t, R_{n-2})]$$
$$+ \frac{\theta \kappa^{1-\theta}}{F(\theta)\Gamma(\theta)} \int_0^t (t-\lambda)^{\theta-1} [P_5(\lambda, R_{n-1}) - P_5(\lambda, R_{n-2})] d\lambda.$$

Notice that:
$$S_n(t) = \sum_{j=0}^{n} \Phi_{1j}(t),$$
$$A_n(t) = \sum_{j=0}^{n} \Phi_{2j}(t),$$
$$I_n(t) = \sum_{j=0}^{n} \Phi_{3j}(t), \qquad (14)$$
$$D_n(t) = \sum_{j=0}^{n} \Phi_{4j}(t),$$
$$R_n(t) = \sum_{j=0}^{n} \Phi_{5j}(t).$$

In the light of Φ_{in} ($i = 1, 2, 3, 4, 5$) definition and benefiting from triangular identity, we obtain:

$$\|\Phi_{1n}(t)\| = \|S_n(t) - S_{n-1}(t)\|$$
$$= \left\| \frac{(1-\theta)\kappa^{1-\theta}}{F(\theta)}[P_1(t, S_{n-1}) - P_1(t, S_{n-2})] \right.$$
$$\left. + \frac{\theta \kappa^{1-\theta}}{F(\theta)\Gamma(\theta)} \int_0^t (t-\lambda)^{\theta-1} [P_1(\lambda, S_{n-1}) - P_1(\lambda, S_{n-2})] d\lambda \right\|. \qquad (15)$$

Since the kernel P_1 provides a Lipschitz condition, we obtain:

$$\|\Phi_{1n}(t)\| = \|S_n(t) - S_{n-1}(t)\|$$
$$\leq \frac{(1-\theta)\kappa^{1-\theta}}{F(\theta)} \varepsilon_1 \|S_{n-1} - S_{n-2}\| + \frac{\theta \kappa^{1-\theta}}{F(\theta)\Gamma(\theta)} \varepsilon_1 \int_0^t (t-\lambda)^{\theta-1} \|S_{n-1} - S_{n-2}\| d\lambda. \qquad (16)$$

and:

$$\|\Phi_{1n}(t)\| \leq \frac{(1-\theta)\kappa^{1-\theta}}{F(\theta)} \varepsilon_1 \|\Phi_{1(n-1)}(t)\| + \frac{\theta \kappa^{1-\theta}}{F(\theta)\Gamma(\theta)} \varepsilon_1 \int_0^t (t-\lambda)^{\theta-1} \|\Phi_{1(n-1)}(\lambda)\| d\lambda. \qquad (17)$$

Analogously, we obtain the following results:

$$\|\Phi_{2n}(t)\| \leq \frac{(1-\theta)\kappa^{1-\theta}}{F(\theta)} \varepsilon_2 \|\Phi_{2(n-1)}(t)\| + \frac{\theta \kappa^{1-\theta}}{F(\theta)\Gamma(\theta)} \varepsilon_2 \int_0^t (t-\lambda)^{\theta-1} \|\Phi_{2(n-1)}(\lambda)\| d\lambda,$$

$$\|\Phi_{3n}(t)\| \leq \frac{(1-\theta)\kappa^{1-\theta}}{F(\theta)} \varepsilon_3 \|\Phi_{3(n-1)}(t)\| + \frac{\theta \kappa^{1-\theta}}{F(\theta)\Gamma(\theta)} \varepsilon_3 \int_0^t (t-\lambda)^{\theta-1} \|\Phi_{3(n-1)}(\lambda)\| d\lambda,$$

$$\|\Phi_{4n}(t)\| \leq \frac{(1-\theta)\kappa^{1-\theta}}{F(\theta)} \varepsilon_4 \|\Phi_{4(n-1)}(t)\| + \frac{\theta \kappa^{1-\theta}}{F(\theta)\Gamma(\theta)} \varepsilon_4 \int_0^t (t-\lambda)^{\theta-1} \|\Phi_{4(n-1)}(\lambda)\| d\lambda$$

$$\|\Phi_{5n}(t)\| \leq \frac{(1-\theta)\kappa^{1-\theta}}{F(\theta)} \varepsilon_5 \|\Phi_{5(n-1)}(t)\| + \frac{\theta \kappa^{1-\theta}}{F(\theta)\Gamma(\theta)} \varepsilon_5 \int_0^t (t-\lambda)^{\theta-1} \|\Phi_{5(n-1)}(\lambda)\| d\lambda. \qquad (18)$$

According to the results in hand, we determine that the system (6) has a solution.

Theorem 2. *The fractional SAIDR system* (6) *has a solution, if there exist* $t_i, i = 1, 2, 3, 4, 5$ *such that:*
$$\frac{(1-\theta)\kappa^{1-\theta}}{F(\theta)}\varepsilon_i + \frac{t_0^\theta \kappa^{1-\theta}}{F(\theta)\Gamma(\theta)}\varepsilon_i < 1.$$

Proof of Theorem 2. We know that $S(t), A(t), I(t), D(t), R(t)$ are bounded functions and the kernels provide a Lipschitz condition. Using Equations (17) and (18), we have:

$$\|\Phi_{1n}(t)\| \leq \|S_n(0)\| \left[\frac{(1-\theta)\kappa^{1-\theta}}{F(\theta)}\varepsilon_1 + \frac{t^\theta \kappa^{1-\theta}}{F(\theta)\Gamma(\theta)}\varepsilon_1\right]^n,$$

$$\|\Phi_{2n}(t)\| \leq \|A_n(0)\| \left[\frac{(1-\theta)\kappa^{1-\theta}}{F(\theta)}\varepsilon_2 + \frac{t^\theta \kappa^{1-\theta}}{F(\theta)\Gamma(\theta)}\varepsilon_2\right]^n$$

$$\|\Phi_{3n}(t)\| \leq \|I_n(0)\| \left[\frac{(1-\theta)\kappa^{1-\theta}}{F(\theta)}\varepsilon_3 + \frac{t^\theta \kappa^{1-\theta}}{F(\theta)\Gamma(\theta)}\varepsilon_3\right]^n \quad (19)$$

$$\|\Phi_{4n}(t)\| \leq \|D_n(0)\| \left[\frac{(1-\theta)\kappa^{1-\theta}}{F(\theta)}\varepsilon_4 + \frac{t^\theta \kappa^{1-\theta}}{F(\theta)\Gamma(\theta)}\varepsilon_4\right]^n,$$

$$\|\Phi_{5n}(t)\| \leq \|R_n(0)\| \left[\frac{(1-\theta)\kappa^{1-\theta}}{F(\theta)}\varepsilon_5 + \frac{t^\theta \kappa^{1-\theta}}{F(\theta)\Gamma(\theta)}\varepsilon_5\right]^n.$$

Thus, Function (14) exists and is smooth. We aim to show that these functions are the solution of Equation (6), assuming that:

$$\begin{aligned}
S(t) - S(0) &= S_n(t) - \overline{g}_{1n}(t), \\
A(t) - A(0) &= A_n(t) - \overline{g}_{2n}(t), \\
I(t) - I(0) &= I_n(t) - \overline{g}_{3n}(t) \\
D(t) - D(0) &= D_n(t) - \overline{g}_{4n}(t), \\
R(t) - R(0) &= R_n(t) - \overline{g}_{5n}(t).
\end{aligned} \quad (20)$$

Thus, we have:

$$\begin{aligned}
\|\overline{g}_{1n}(t)\| &= \left\|\frac{(1-\theta)\kappa^{1-\theta}}{F(\theta)}[P_1(t,S) - P_1(t,S_{n-1})] \right. \\
&\quad \left. + \frac{\theta\kappa^{1-\theta}}{F(\theta)\Gamma(\theta)}\int_0^t (t-\lambda)^{\theta-1}[P_1(\lambda,S) - P_1(\lambda, S_{n-1})]d\lambda\right\| \\
&\leq \frac{(1-\theta)\kappa^{1-\theta}}{F(\theta)}\|P_1(t,S) - P_1(t,S_{n-1})\| \quad (21) \\
&\quad + \frac{\theta\kappa^{1-\theta}}{F(\theta)\Gamma(\theta)}\int_0^t (t-\lambda)^{\theta-1}\|P_1(\lambda,S) - P_1(\lambda, S_{n-1})d\lambda\| \\
&\leq \frac{(1-\theta)\kappa^{1-\theta}}{F(\theta)}\varepsilon_1\|S - S_{n-1}\| + \frac{t^\theta \kappa^{1-\theta}}{F(\theta)\Gamma(\theta)}\varepsilon_1\|S - S_{n-1}\|.
\end{aligned}$$

Repeating this method, we obtain at t_0:

$$\|\overline{g}_{1n}(t)\| \leq \left(\frac{(1-\theta)\kappa^{1-\theta}}{F(\theta)} + \frac{t_0^\theta \kappa^{1-\theta}}{F(\theta)\Gamma(\theta)}\right)^{n+1} \gamma_1^{n+1} M. \quad (22)$$

As n approaches ∞, $\|\overline{g}_{1n}(t)\| \to 0$. In the same way, it can be shown that $\|\overline{g}_{in}(t)\| \to 0$ ($i = 2, 3, 4, 5$). □

To show the uniqueness of the solution, we suppose that the system (6) has another solution $S_1(t), A_1(t), I_1(t), R_1(t)$ then:

$$\|S(t) - S_1(t)\| = \left\| \frac{(1-\theta)\kappa^{1-\theta}}{F(\theta)} [P_1(t,S) - P_1(t,S_1)] \right.$$

$$\left. + \frac{\theta \kappa^{1-\theta}}{F(\theta)\Gamma(\theta)} \int_0^t (t-\lambda)^{\theta-1} [P_1(\lambda,S) - P_1(\lambda,S_1)] d\lambda \right\|$$

$$\leq \frac{(1-\theta)\kappa^{1-\theta}}{F(\theta)} \|P_1(t,S) - P_1(t,S_1)\|$$

$$+ \frac{\theta \kappa^{1-\theta}}{F(\theta)\Gamma(\theta)} \int_0^t (t-\lambda)^{\theta-1} \|P_1(\lambda,S) - P_1(\lambda,S_1)\| d\lambda. \quad (23)$$

Regarding the Lipschitz condition of S, we gain:

$$\|S(t) - S_1(t)\| \leq \frac{(1-\theta)\kappa^{1-\theta}}{F(\theta)} \varepsilon_1 \|S(t) - S_1(t)\| + \frac{t^\theta \kappa^{1-\theta}}{F(\theta)\Gamma(\theta)} \varepsilon_1 \|S(t) - S_1(t)\|. \quad (24)$$

This gives:

$$\|S(t) - S_1(t)\| \left(1 - \frac{(1-\theta)\kappa^{1-\theta}}{F(\theta)} \varepsilon_1 - \frac{t^\theta \kappa^{1-\theta}}{F(\theta)\Gamma(\theta)} \varepsilon_1 \right) \leq 0. \quad (25)$$

Obviously $S(t) = S_1(t)$, if the following inequality holds:

$$\left(1 - \frac{(1-\theta)\kappa^{1-\theta}}{F(\theta)} \varepsilon_1 - \frac{t^\theta \kappa^{1-\theta}}{F(\theta)\Gamma(\theta)} \varepsilon_1 \right) > 0$$

then $\|S(t) - S_1(t)\| = 0$. Therefore, we gain:

$$S(t) = S_1(t).$$

In the same way, we find:

$$A(t) = A_1(t), I(t) = I_1(t), D(t) = D_1(t), R(t) = R_1(t).$$

4. Stability Analysis by Fixed Point Theory

In this section, we give a special solution of the fractional SAIDR model (6) with a recursive formula by using Laplace transform. The Laplace transform for the AB fractional derivative was introduced by Atangana and Baleanu [22] as follows:

Theorem 3. *Let $\theta \in [0,1]$, $a < b$ and $g \in H^1(a,b)$. The Laplace transform for the AB derivative in the Caputo type is presented by:*

$$L\left\{{}^{ABC}_0 D^\theta_t [g(t)]\right\}(p) = \frac{F(\theta)}{1-\theta} \frac{p^\theta L\{g(t)\}(p) - p^{\theta-1} g(0)}{p^\theta + \frac{\theta}{1-\theta}}. \quad (26)$$

We apply the Laplace transform to the Equation (6), then:

$$L\left(\frac{1}{\kappa^{1-\theta}} {}^{ABC}_{0}D^\theta_t S(t)\right)(p) = L(\mu N - \gamma S(t) - \beta S(t)I(t) - \mu S(t))(p),$$

$$L\left(\frac{1}{\kappa^{1-\theta}} {}^{ABC}_{0}D^\theta_t A(t)\right)(p) = L(\beta S(t)I(t) - \delta A(t) - \eta A(t) - \mu A(t))(p),$$

$$L\left(\frac{1}{\kappa^{1-\theta}} {}^{ABC}_{0}D^\theta_t I(t)\right)(p) = L(\eta A(t) - \sigma I(t) - \tau I(t) - \mu I(t))(p),$$

$$L\left(\frac{1}{\kappa^{1-\theta}} {}^{ABC}_{0}D^\theta_t D(t)\right)(p) = L(\tau I(t) - \varphi D(t) - \mu D(t))(p),$$

$$L\left(\frac{1}{\kappa^{1-\theta}} {}^{ABC}_{0}D^\theta_t R(t)\right)(p) = L(\gamma S(t) + \delta A(t) + \sigma I(t) + \varphi D(t) - \mu R(t))(p).$$

Benefiting from the Laplace transform definition of the AB derivative, we obtain:

$$\frac{F(\theta)}{1-\theta} \frac{1}{p^\theta + \frac{\theta}{1-\theta}} \left(p^\theta L(S(t))(p) - p^{\theta-1} S(0)\right)$$
$$= \kappa^{1-\theta} L(\mu N - \gamma S(t) - \beta S(t)I(t) - \mu S(t))(p),$$

$$\frac{F(\theta)}{1-\theta} \frac{1}{p^\theta + \frac{\theta}{1-\theta}} \left(p^\theta L(A(t))(p) - p^{\theta-1} A(0)\right)$$
$$= \kappa^{1-\theta} L(\beta S(t)I(t) - \delta A(t) - \eta A(t) - \mu A(t))(p),$$

$$\frac{F(\theta)}{1-\theta} \frac{1}{p^\theta + \frac{\theta}{1-\theta}} \left(p^\theta L(I(t))(p) - p^{\theta-1} I(0)\right)$$
$$= \kappa^{1-\theta} L(\eta A(t) - \sigma I(t) - \tau I(t) - \mu I(t))(p),$$

$$\frac{F(\theta)}{1-\theta} \frac{1}{p^\theta + \frac{\theta}{1-\theta}} \left(p^\theta L(D(t))(p) - p^{\theta-1} D(0)\right)$$
$$= \kappa^{1-\theta} L(\tau I(t) - \varphi D(t) - \mu D(t))(p),$$

$$\frac{F(\theta)}{1-\theta} \frac{1}{p^\theta + \frac{\theta}{1-\theta}} \left(p^\theta L(R(t))(p) - p^{\theta-1} R(0)\right)$$
$$= \kappa^{1-\theta} L(\gamma S(t) + \delta A(t) + \sigma I(t) + \varphi D(t) - \mu R(t))(p). \tag{27}$$

Regulating Equation (27), we derive:

$$L(S(t))(p) = \frac{1}{p}S(0) + \psi\kappa^{1-\theta} \times L(\mu N - \gamma S(t) - \beta S(t)I(t) - \mu S(t))(p),$$

$$L(A(t))(p) = \frac{1}{p}A(0) + \psi\kappa^{1-\theta} \times L(\beta S(t)I(t) - \delta A(t) - \eta A(t) - \mu A(t))(p),$$

$$L(I(t))(p) = \frac{1}{p}I(0) + \psi\kappa^{1-\theta} \times L(\eta A(t) - \sigma I(t) - \tau I(t) - \mu I(t))(p),$$

$$L(D(t))(p) = \frac{1}{p}D(0) + \psi\kappa^{1-\theta} \times L(\tau I(t) - \varphi D(t) - \mu D(t))(p),$$

$$L(R(t))(p) = \frac{1}{p}R(0) + \psi\kappa^{1-\theta} \times L(\gamma S(t) + \delta A(t) + \sigma I(t) + \varphi D(t) - \mu R(t))(p),$$

where:

$$\psi = \left(1 - \theta + \frac{\theta}{p^\theta}\right)\frac{1}{F(\theta)}.$$

Thus, we have the following iterative formula by taking the inverse Laplace transform both sides of all equations, as follows:

$$S_{n+1}(t) = S_n(0) + L^{-1}\left(\psi\kappa^{1-\theta}ST(\mu N - \gamma S(t) - \beta S(t)I(t) - \mu S(t))(p)\right),$$

$$A_{n+1}(t) = A_n(0) + L^{-1}\left(\psi\kappa^{1-\theta}ST(\beta S(t)I(t) - \delta A(t) - \eta A(t) - \mu A(t))(p)\right),$$

$$I_{n+1}(t) = I_n(0) + L^{-1}\left(\psi\kappa^{1-\theta}ST(\eta A(t) - \sigma I(t) - \tau I(t) - \mu I(t))(p)\right), \qquad (28)$$

$$D_{n+1}(t) = D_n(0) + L^{-1}\left(\psi\kappa^{1-\theta}ST(\tau I(t) - \varphi D(t) - \mu D(t))(p)\right),$$

$$R_{n+1}(t) = R_n(0) + L^{-1}\left(\psi\kappa^{1-\theta}ST(\gamma S(t) + \delta A(t) + \sigma I(t) + \varphi D(t) - \mu R(t))(p)\right).$$

The approximate solution of the model (6) is as below:

$$S(t) = \lim_{n\to\infty} S_n(t), \ A(t) = \lim_{n\to\infty} A_n(t), \ I(t) = \lim_{n\to\infty} I_n(t),$$

$$D(t) = \lim_{n\to\infty} D_n(t), \ R(t) = \lim_{n\to\infty} R_n(t).$$

Stability Analysis of Iteration Method

Considering the Banach space $(X, \|.\|)$, a self map T on X and the recursive method $q_{n+1} = \phi(T, q_n)$. We assume that $\{t_n\} \subset \gamma$ is the fixed point set of T which $\gamma(T) \neq \varnothing$ and $\lim_{n\to\infty} q_n = q \in \gamma(t)$. We also suppose that $\{t_n\} \subset \gamma$ and $r_n = \|t_{n+1} - \phi(T, t_n)\|$. If $\lim_{n\to\infty} r_n = 0$ implies that $\lim_{n\to\infty} t_n = q$, then the iteration method $q_{n+1} = \phi(T, q_n)$ is T-stable. We suppose that our sequence $\{t_n\}$ has an upper boundary. If Picard's iteration $q_{n+1} = Tq_n$ satisfies all conditions, then $q_{n+1} = Tq_n$ is T-stable.

Theorem 4. *Let $(X, \|.\|)$ be Banach space and $T : X \to X$ be a map satisfying:*

$$\|T_x - T_y\| \leq K\|x - T_x\| + k\|x - y\|,$$

for all $x, y \in X$, where $0 \leq K$, $0 \leq k < 1$. Then, T is Picard T-stable [32].

Theorem 5. *Assume that T is a self map defined as below:*

$$T(S_n(t)) = S_{n+1}(t)$$
$$= S_n(t) + L^{-1}\Big(\psi\kappa^{1-\theta} \times L(\mu N - \gamma S(t) - \beta S(t)I(t) - \mu S(t))(p)\Big),$$
$$T(A_n(t)) = A_{n+1}(t)$$
$$= A_n(t) + L^{-1}\Big(\psi\kappa^{1-\theta} \times L(\beta S(t)I(t) - \delta A(t) - \eta A(t) - \mu A(t))(p)\Big),$$
$$T(I_n(t)) = I_{n+1}(t)$$
$$= I_n(t) + L^{-1}\Big(\psi\kappa^{1-\theta} \times L(\eta A(t) - \sigma I(t) - \tau I(t) - \mu I(t))(p)\Big),$$
$$T(D_n(t)) = D_{n+1}(t)$$
$$= D_n(t) + L^{-1}\Big(\psi\kappa^{1-\theta} \times L(\tau I(t) - \varphi D(t) - \mu D(t))(p)\Big),$$
$$T(R_n(t)) = R_{n+1}(t)$$
$$= R_n(t) + L^{-1}\Big(\psi\kappa^{1-\theta} \times L(\gamma S(t) + \delta A(t) + \sigma I(t) + \varphi D(t) - \mu R(t))(p)\Big).$$

Then, the iteration is T-stable in $L^1(a,b)$ if the following statements are achieved:

$$1 - (\mu + \gamma)h_1(\overline{\gamma}) - \beta(M_3 + M_1)h_2(\overline{\gamma}) < 1,$$
$$1 - \beta(M_3 + M_1)h_3(\overline{\gamma}) - (\delta + \eta + \mu)h_4(\overline{\gamma}) < 1,$$
$$1 + \eta h_5(\overline{\gamma}) - (\sigma + \tau + \mu)h_6(\overline{\gamma}) < 1,$$
$$1 + \tau h_7(\overline{\gamma}) - (\varphi + \mu)h_8(\overline{\gamma}) < 1,$$
$$1 + \gamma h_9(\overline{\gamma}) + \delta h_{10}(\overline{\gamma}) + \sigma h_{11}(\overline{\gamma}) + \varphi h_{12}(\overline{\gamma}) - \mu h_{13}(\overline{\gamma}) < 1.$$

Proof. To show that T has a fixed point, we evaluated the following for $(i,j) \in \mathbb{N} \times \mathbb{N}$:

$$T(S_i(t)) - T(S_j(t)) = S_i(t) - S_j(t)$$
$$+ L^{-1}\Big(\psi\kappa^{1-\theta} \times L(\mu N - \gamma S_i(t) - \beta S_i(t)I_i(t) - \mu S_i(t))(p)\Big) \quad (29)$$
$$- L^{-1}\Big(\psi\kappa^{1-\theta} \times L(\mu N - \gamma S_j(t) - \beta S_j(t)I_j(t) - \mu S_j(t))(p)\Big).$$

Taking the norm, Equation (30) is converted to:

$$\|T(S_i(t)) - T(S_j(t))\| = \Big\| S_i(t) - S_j(t)$$
$$+ L^{-1}\Big(\psi\kappa^{1-\theta} \times L(\mu N - \gamma S_i(t) - \beta S_i(t)I_i(t) - \mu S_i(t))(p)\Big) \quad (30)$$
$$- L^{-1}\Big(\psi\kappa^{1-\theta} \times L(\mu N - \gamma S_j(t) - \beta S_j(t)I_j(t) - \mu S_j(t))(p)\Big)\Big\|.$$

Using norm properties, we obtain:

$$\|T(S_i(t)) - T(S_j(t))\| \leq \|S_i(t) - S_j(t)\|$$
$$+ L^{-1}\Big(\psi\kappa^{1-\theta} \times L\Big(\begin{array}{c}\|-(\gamma+\mu)(S_i(t) - S_j(t))\\ -\beta(I_i(t)(S_i(t) - S_j(t)) + S_j(t)(I_i(t) - I_j(t)))\|\end{array}\Big)(p)\Big) \quad (31)$$

Since the solutions play the same role, we assume that:

$$\|S_i(t) - S_j(t)\| \cong \|A_i(t) - A_j(t)\|$$
$$\cong \|I_i(t) - I_j(t)\| \cong \|D_i(t) - D_j(t)\| \cong \|R_i(t) - R_j(t)\| \quad (32)$$

From Equations (31) and (32), we find:

$$\|T(S_i(t)) - T(S_j(t))\| \leq \|S_i(t) - S_j(t)\|$$
$$+ L^{-1}\left(\psi\kappa^{1-\theta} \times L\left(\|-(\gamma+\mu)(S_i(t) - S_j(t))\|\right)(p)\right) \qquad (33)$$
$$+ L^{-1}\left(\psi\kappa^{1-\theta} \times L\left(\|-\beta(I_i(S_i(t) - S_j(t)) + S_j(S_i(t) - S_j(t)))\|\right)(p)\right)$$

Because $S_i(t)$, $A_i(t)$, $I_i(t)$, $D_i(t)$ and $R_i(t)$ are bounded, for all t there exists M_i, $i = 1, 2, 3, 4, 5$ such that:

$$\|S_i(t)\| \leq M_1, \|A_i(t)\| \leq M_2,$$
$$\|I_i(t)\| \leq M_3, \|D_i(t)\| \leq M_4, \|R_i(t)\| \leq M_5. \qquad (34)$$

Here, considering Equations (33) and (34), we have:

$$\|T(S_i(t)) - T(S_j(t))\| \leq \|S_i(t) - S_j(t)\| \times [1 - (\gamma+\mu)h_1(\overline{\gamma}) - \beta(M_3 + M_1)h_2(\overline{\gamma})] \qquad (35)$$

where h_i are functions from $L^{-1}\{\psi\kappa^{1-\theta}L\}$. In an analogous way, we achieve:

$$\|T(A_i(t)) - T(A_j(t))\| \leq \|A_i(t) - A_j(t)\|$$
$$\times [1 - \beta(M_3 + M_1)h_3(\overline{\gamma}) - (\delta + \eta + \mu)h_4(\overline{\gamma})],$$

$$\|T(I_i(t)) - T(I_j(t))\| \leq \|I_i(t) - I_j(t)\|$$
$$\times [1 + \eta h_5(\overline{\gamma}) - (\sigma + \tau + \mu)h_6(\overline{\gamma})],$$

$$\|T(D_i(t)) - T(D_j(t))\| \leq \|D_i(t) - D_j(t)\|$$
$$\times [1 + \tau h_7(\overline{\gamma}) - (\varphi + \mu)h_8(\overline{\gamma})],$$

$$\|T(R_i(t)) - T(R_j(t))\| \leq \|R_i(t) - R_j(t)\|$$
$$\times [1 + \gamma h_9(\overline{\gamma}) + \delta h_{10}(\overline{\gamma}) + \sigma h_{11}(\overline{\gamma}) + \varphi h_{12}(\overline{\gamma}) - \mu h_{13}(\overline{\gamma})] \qquad (36)$$

where:

$$1 - (\gamma + \mu)h_1(\overline{\gamma}) - \beta(M_3 + M_1)h_2(\overline{\gamma}) < 1,$$
$$1 - \beta(M_3 + M_1)h_3(\overline{\gamma}) - (\delta + \eta + \mu)h_4(\overline{\gamma}) < 1,$$
$$1 + \eta h_5(\overline{\gamma}) - (\sigma + \tau + \mu)h_6(\overline{\gamma}) < 1,$$
$$1 + \tau h_7(\overline{\gamma}) - (\varphi + \mu)h_8(\overline{\gamma}) < 1,$$
$$1 + \gamma h_9(\overline{\gamma}) + \delta h_{10}(\overline{\gamma}) + \sigma h_{11}(\overline{\gamma}) + \varphi h_{12}(\overline{\gamma}) - \mu h_{13}(\overline{\gamma}) < 1.$$

Therefore, T has a fixed point. Considering Equations (35) and (36), we assume:

$$k = (0, 0, 0, 0, 0),$$
$$K = \left\{ \begin{array}{l} 1 - (\gamma + \mu)h_1(\overline{\gamma}) - \beta(M_3 + M_1)h_2(\overline{\gamma}), \\ 1 - \beta(M_3 + M_1)h_3(\overline{\gamma}) - (\delta + \eta + \mu)h_4(\overline{\gamma}), \\ 1 + \eta h_5(\overline{\gamma}) - (\sigma + \tau + \mu)h_6(\overline{\gamma}), \\ 1 + \tau h_7(\overline{\gamma}) - (\varphi + \mu)h_8(\overline{\gamma}) \\ 1 + \gamma h_9(\overline{\gamma}) + \delta h_{10}(\overline{\gamma}) + \sigma h_{11}(\overline{\gamma}) + \varphi h_{12}(\overline{\gamma}) - \mu h_{13}(\overline{\gamma}) \end{array} \right\}.$$

Thus, all the conditions of Theorem 4 are satisfied. This completes the proof. □

5. Numerical Results

With the aim of obtaining the solution, through some equations of fractional derivatives with a non-local and non-singular kernel, Toufik and Atangana [33] presented a novel numerical scheme based on the fundamental theorem of fractional calculus and a two-step Lagrange polynomial. We give the so-called method for the fractional SMS-based worm propagation model in mobile networks (6). At a point $t = t_{n+1}$, we apply this scheme to Equation (12):

$$\begin{aligned}
S_{n+1} = S_0 &+ \frac{(1-\theta)\kappa^{1-\theta}}{F(\theta)} P_1(t_n, S(t_n)) \\
&+ \frac{\theta}{F(\theta)} \sum_{k=0}^{n} \left(\frac{h^\theta P_1(t_k, S_k)}{\Gamma(\theta+2)} \left((n-k+1)^\theta (n-k+2+\theta) - (n-k)^\theta (n-k+2+2\theta) \right) \right. \\
&\left. - \frac{h^\theta P_1(t_{k-1}, S_{k-1})}{\Gamma(\eta+2)} \left((n-k+1)^{\theta+1} - (n-k)^\theta (n-k+1+\theta) \right) +^1 L_n^\theta,
\end{aligned} \quad (37)$$

$$\begin{aligned}
A_{n+1} = A_0 &+ \frac{(1-\theta)\kappa^{1-\theta}}{F(\theta)} P_2(t_n, A(t_n)) \\
&+ \frac{\theta}{F(\theta)} \sum_{k=0}^{n} \left(\frac{h^\theta P_2(t_k, A_k)}{\Gamma(\theta+2)} \left((n-k+1)^\theta (n-k+2+\theta) - (n-k)^\theta (n-k+2+2\theta) \right) \right. \\
&\left. - \frac{h^\theta P_2(t_{k-1}, A_{k-1})}{\Gamma(\eta+2)} \left((n-k+1)^{\theta+1} - (n-k)^\theta (n-k+1+\theta) \right) +^2 L_n^\theta,
\end{aligned} \quad (38)$$

$$\begin{aligned}
I_{n+1} = I_0 &+ \frac{(1-\theta)\kappa^{1-\theta}}{F(\theta)} P_3(t_n, I(t_n)) \\
&+ \frac{\theta}{F(\theta)} \sum_{k=0}^{n} \left(\frac{h^\theta P_3(t_k, I_k)}{\Gamma(\theta+2)} \left((n-k+1)^\theta (n-k+2+\theta) - (n-k)^\theta (n-k+2+2\theta) \right) \right. \\
&\left. - \frac{h^\theta P_3(t_{k-1}, I_{k-1})}{\Gamma(\eta+2)} \left((n-k+1)^{\theta+1} - (n-k)^\theta (n-k+1+\theta) \right) +^3 L_n^\theta,
\end{aligned} \quad (39)$$

$$\begin{aligned}
D_{n+1} = D_0 &+ \frac{(1-\theta)\kappa^{1-\theta}}{F(\theta)} P_4(t_n, D(t_n)) \\
&+ \frac{\theta}{F(\theta)} \sum_{k=0}^{n} \left(\frac{h^\theta P_4(t_k, D_k)}{\Gamma(\theta+2)} \left((n-k+1)^\theta (n-k+2+\theta) - (n-k)^\theta (n-k+2+2\theta) \right) \right. \\
&\left. - \frac{h^\theta P_4(t_{k-1}, D_{k-1})}{\Gamma(\eta+2)} \left((n-k+1)^{\theta+1} - (n-k)^\theta (n-k+1+\theta) \right) +^4 L_n^\theta,
\end{aligned} \quad (40)$$

$$\begin{aligned}
R_{n+1} = R_0 &+ \frac{(1-\theta)\kappa^{1-\theta}}{F(\theta)} P_5(t_n, R(t_n)) \\
&+ \frac{\theta}{F(\theta)} \sum_{k=0}^{n} \left(\frac{h^\theta P_5(t_k, R_k)}{\Gamma(\theta+2)} \left((n-k+1)^\theta (n-k+2+\theta) - (n-k)^\theta (n-k+2+2\theta) \right) \right. \\
&\left. - \frac{h^\theta P_5(t_{k-1}, R_{k-1})}{\Gamma(\eta+2)} \left((n-k+1)^{\theta+1} - (n-k)^\theta (n-k+1+\theta) \right) +^5 L_n^\theta,
\end{aligned} \quad (41)$$

where $^iL_n^\theta$, $i = 1, 2, 3, 4, 5$ are remainder terms given by

$$^1L_n^\theta = \frac{\theta}{F(\theta)\Gamma(\theta)} \sum_{k=0}^{n} \int_{t_k}^{t_{k-1}} \frac{(\lambda - t_k)(\lambda - t_{k-1})}{2!} \frac{\partial^2}{\partial \lambda^2} [P_1(\lambda, S(\lambda))]_{\lambda=\varepsilon_\lambda} (t_{n+1} - \lambda)^{\eta-1} d\lambda,$$

$$^2L_n^\theta = \frac{\theta}{F(\theta)\Gamma(\theta)} \sum_{k=0}^{n} \int_{t_k}^{t_{k-1}} \frac{(\lambda - t_k)(\lambda - t_{k-1})}{2!} \frac{\partial^2}{\partial \lambda^2} [P_2(\lambda, A(\lambda))]_{\lambda=\varepsilon_\lambda} (t_{n+1} - \lambda)^{\eta-1} d\lambda,$$

$$^3L_n^\theta = \frac{\theta}{F(\theta)\Gamma(\theta)} \sum_{k=0}^{n} \int_{t_k}^{t_{k-1}} \frac{(\lambda - t_k)(\lambda - t_{k-1})}{2!} \frac{\partial^2}{\partial \lambda^2} [P_3(\lambda, I(\lambda))]_{\lambda=\varepsilon_\lambda} (t_{n+1} - \lambda)^{\eta-1} d\lambda,$$

$$^4L_n^\theta = \frac{\theta}{F(\theta)\Gamma(\theta)} \sum_{k=0}^{n} \int_{t_k}^{t_{k-1}} \frac{(\lambda - t_k)(\lambda - t_{k-1})}{2!} \frac{\partial^2}{\partial \lambda^2} [P_4(\lambda, D(\lambda))]_{\lambda=\varepsilon_\lambda} (t_{n+1} - \lambda)^{\eta-1} d\lambda,$$

$$^5L_n^\theta = \frac{\theta}{F(\theta)\Gamma(\theta)} \sum_{k=0}^{n} \int_{t_k}^{t_{k-1}} \frac{(\lambda - t_k)(\lambda - t_{k-1})}{2!} \frac{\partial^2}{\partial \lambda^2} [P_5(\lambda, R(\lambda))]_{\lambda=\varepsilon_\lambda} (t_{n+1} - \lambda)^{\eta-1} d\lambda.$$

The numerical productions of the model (6) are hereby displayed by the foregoing method. To this end, the initial conditions are posited as $S(0) = 99,000$, $A(0) = 500$, $I(0) = 500$, $D(0) = 0$, $R(0) = 0$ and the variable factors $\mu = 0.000001$, $\eta = 0.003$, $\delta = 0.003$, $\sigma = 0.004$, $\tau = 0.001$, $\varphi = 0.007$, $\beta = 0.000003$ are selected as specified in [29]. Any elevation or decline in the susceptible nodes, affected nodes, infected nodes, suspended nodes, or recovered nodes with respect to the distinct fractional order, and the numerical amounts of the chosen variable quantities are displayed by the figures. Figure 1a illustrates the preliminary elevation in the quantity of infected nodes which rises to the highest point, nearly 20% of the overall amount present within the structure at approximately the 300th minute; but afterwards, this number diminishes at a fast pace. It follows from here that SMS is one of the ways that enables the worm to be quickly transmitted throughout the mobile network. Figure 1b–d indicate that the quantity of infected nodes gradually rises while the fractional order declines. Hence, we can assert that SMS is a means for the worm to transmit itself gradually; however, the quantity of infected nodes that continues existing in the system is higher.

In order to illustrate the reasonableness of the fractional SAIDR model, let us herein investigate the ratio of infection and the ratio of transition between infected and suspended states, which are two substantial variables. Figure 2a,b evince that a larger quantity of nodes will be contaminated more quickly as the infection rate scales up. To put it differently, as β for the fractional order $\theta = 0.95$ and $\theta = 0.65$ elevates, the worm circulates more swiftly. For this reason, it is possible to say that reducing the infection ratio results in an acceleration of the duration during which the harmful software is wiped out. It follows from Figure 3a,b that the graph $\tau = 0.003$ hits the lowest point when the graph $\tau = 0.001$ comes to the topmost more or less simultaneously, approximately at the 400th min, reciprocally involving the three graphs' highest points. A smaller amount of nodes becomes simultaneously contaminated as a consequence of the escalation in transition ratios to the suspended state from the infected state. The reason is that a smaller number of nodes continue staying in the infected state since a greater amount of nodes are able to shift to the suspended state from the infected state with the escalating ratio.

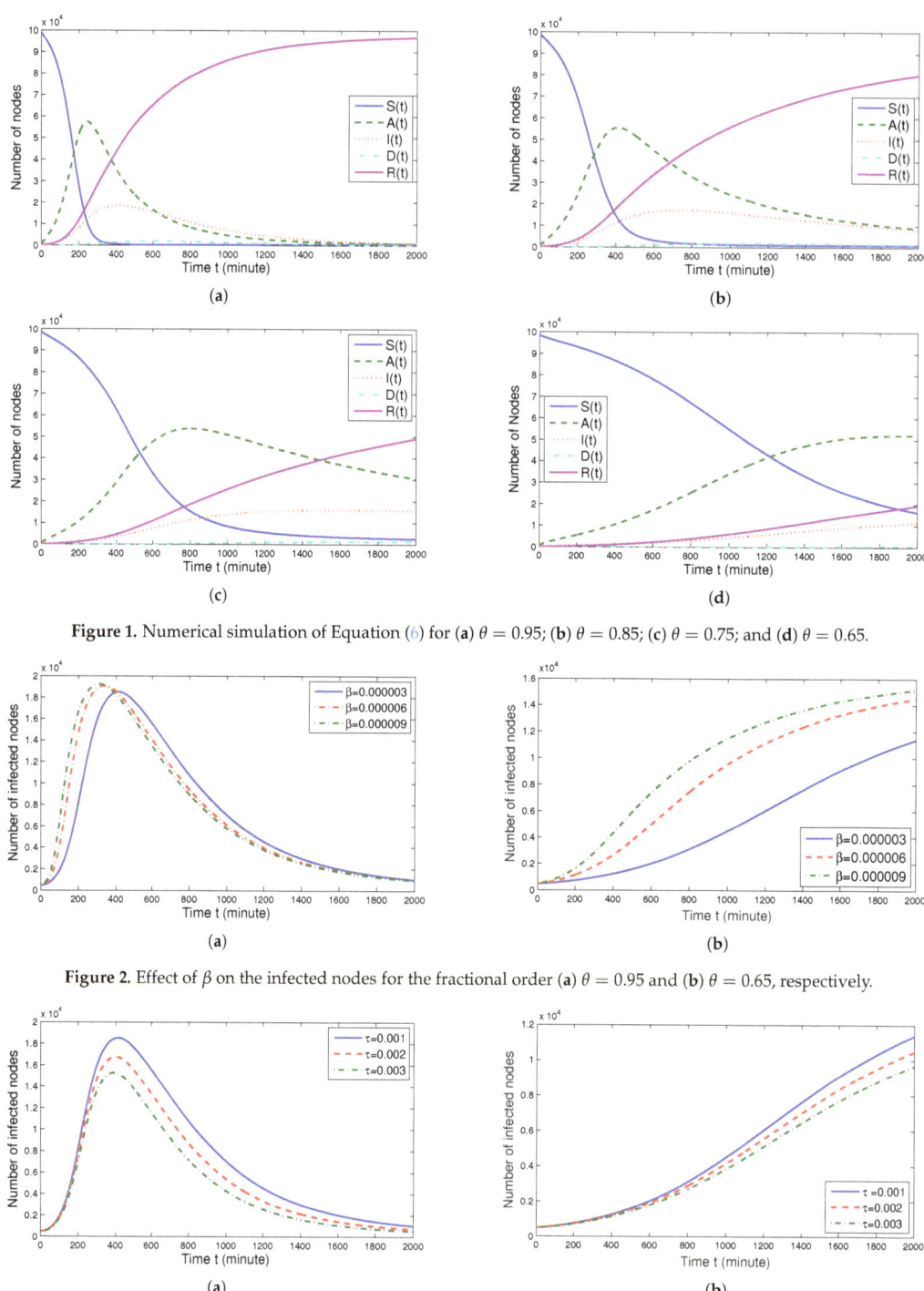

Figure 1. Numerical simulation of Equation (6) for (**a**) $\theta = 0.95$; (**b**) $\theta = 0.85$; (**c**) $\theta = 0.75$; and (**d**) $\theta = 0.65$.

Figure 2. Effect of β on the infected nodes for the fractional order (**a**) $\theta = 0.95$ and (**b**) $\theta = 0.65$, respectively.

Figure 3. Relation between τ and infected nodes for the fractional order (**a**) $\theta = 0.95$ and (**b**) $\theta = 0.65$, respectively.

6. Conclusions

A novel fractional order derivative involving a Mittag–Leffler kernel has recently been introduced by Atangana in cooperation with Baleanu. First and foremost, the AB derivative broadens the scope of the model grounding on [29] so that we can consider the additional implementation of the relevant fractional derivative and monitor the propagation of computer worms in mobile networks more comprehensively. We propound a fractional model which carries the probability of not having any closed form solution since it is nonlinear. Hence, the circumstances providing the existence and uniqueness of the solution regarding this fractional SAIDR model become evident, and the special solution is thus reproduced through the Laplace transform. Lastly, we apply numerical simulations of this model so as to reach efficacy with this novel derivative provided with a fractional order. Additionally, we express the impact that infection ratio has on infected nodes numerically, and on the grounds of the relevant graphics, we conclude that diminishing the ratio of infection speeds up the duration during which the malignant software are eliminated.

Author Contributions: Conceptualization, E.U. and S.U.; methodology, S.U., F.E. and N.Ö.; software, F.E.; validation, E.U., S.U., F.E. and N.Ö.; formal analysis, F.E. and N.Ö.; investigation, E.U., S.U., F.E. and N.Ö.; resources, E.U., S.U., F.E. and N.Ö.; writing—original draft preparation, E.U., S.U., F.E. and N.Ö.; writing—review and editing, E.U., S.U., F.E. and N.Ö.; visualization, S.U. and F.E.; supervision, N.Ö.; project administration, S.U. All authors have read and agreed to the published version of the manuscript.

Funding: This research is supported by Balikesir University Research Grant No. BAP 2020/014.

Data Availability Statement: Not applicable.

Conflicts of Interest: The authors declare no conflict of interest.

References

1. Abraham, S.; Smith, I.C. An overview of social engineering malware: Trends, tactics, and implications. *Technol. Soc.* **2010**, *32*, 183–196. [CrossRef]
2. CNCENT/CC, CCKUN-A Mobile Malware Spreading in Social Relationship Networks by SMS. 2013. Available online: https://www.cert.org.cn/publish/main/8/2013/20130924145326642925406/20130924145326642925406_.html (accessed on 10 February 2021).
3. Computer World, Android SMS Worm Selfmite Is Back, More Aggressive Than Ever. 2013. Available online: http://www.computerworld.com/article/2824619/android-sms-worm-selfmite-is-back-more-aggressive-than-ever.html (accessed on 10 February 2021).
4. CNCENT/CC. The bulletin about the outbreak and response of the xxShenQi malware. 2014. Available online: https://www.cert.org.cn/publish/main/12/2014/20140803174220396365334/20140803174220396365334_.html (accessed on 10 February 2021).
5. Podlubny, I. *Fractional Differential Equations: An Introduction to Fractional Derivatives to Methods of Their Solution and Some of Their Applications*; Academic Press: San Diego, CA, USA, 1999.
6. Caputo, M.; Mainardi, F. A new dissipation model based on memory mechanism. *Pure Appl. Geophys.* **1971**, *91*, 134–147. [CrossRef]
7. Caputo, M.; Fabrizio, F. A new definition of fractional derivative without singular kerne. *Prog. Fract. Differ. Appl.* **2015**, *1*, 73–85.
8. Koca, I. Analysis of rubella disease model with non-local and non-singular fractional derivatives. *Int. J. Optim. Control. Theor. Appl.* **2018**, *8*, 17–25. [CrossRef]
9. Jajarmi, A.; Baleanu, D. A new fractional analysis on the interaction of HIV with CD4+ T-cells. *Chaos Solitons Fractals* **2018**, *113*, 221–229. [CrossRef]
10. Uçar, S.; Özdemir, N.; Koca, I.; Altun, E. Novel analysis of the fractional glucose insulin regulatory system with non-singular kernel derivative. *Eur. Phys. J. Plus* **2020**, *135*, 414. [CrossRef]
11. Özdemir, N.; Agrawal, O.P.; İskender, B.B.; Karadeniz, D. Fractional optimal control of a 2-dimensional distributed system using eigenfunctions. *Nonlinear Dyn.* **2009**, *55*, 251–260. [CrossRef]
12. Baleanu, D.; Fernandez, A.; Akgül, A. On a fractional operator combining proportional and classical differintegrals. *Mathematics* **2020**, *8*, 13. [CrossRef]
13. Uçar, E.; Özdemir, N.; Altun, E. Fractional order model of immune cells influenced by cancer cells. *Math. Model. Nat.* **2019**, *14*, 12.
14. Evirgen, F.; Özdemir, N. Multistage adomian decomposition method for solving NLP problems over a nonlinear fractional dynamical system. *J. Comput. Nonlinear Dyn.* **2011**, *6*. [CrossRef]
15. Evirgen, F. Conformable Fractional Gradient Based Dynamic System for Constrained Optimization Problem. *Acta Phys. Pol. A* **2017**, *132*, 1066–1069. [CrossRef]

16. Uçar, E.; Özdemir, N. A fractional model of cancer-immune system with Caputo and Caputo–Fabrizio derivatives. *Eur. Phys. J. Plus* **2021**, *136*, 1–17. [CrossRef] [PubMed]
17. Aljoudi, S.; Ahmad, B.; Alsaedi, A. Existence and uniqueness results for a coupled system of Caputo-Hadamard fractional differential equations with nonlocal Hadamard type integral boundary conditions. *Fractal Fract.* **2020**, *4*, 15. [CrossRef]
18. Baleanu, D.; Hakimeh M.; Shahram, R. A fractional differential equation model for the COVID-19 transmission by using the Caputo–Fabrizio derivative. *Adv. Differ. Eq.* **2020**, *2020*, 299. [CrossRef] [PubMed]
19. Aliyu, A. I.; Inc M.; Yusuf, A.; Baleanu, D. A fractional model of vertical transmission and cure of vector-borne diseases pertaining to the Atangana–Baleanu fractional derivatives. *Chaos Solitons Fractals* **2018**, *116*, 268–277. [CrossRef]
20. Uçar, S. Analysis of a basic SEIRA model with Atangana-Baleanu derivative. *AIMS Math.* **2020**, *5*, 1411–1424. [CrossRef]
21. Kumar, D.; Singh, J. New aspects of fractional epidemiological model for computer viruses with Mittag–Leffler law. In *Mathematical Modelling in Health, Social and Applied Sciences*; Dutta, H., Ed.; Springer: Singapore, 2020; pp. 283–301.
22. Atangana, A.; Baleanu, D. New fractional derivatives with nonlocal and non-singular kernel: Theory and application to heat transfer model. *Therm. Sci.* **2016**, *20*, 763–769. [CrossRef]
23. Fernandez, A.; Baleanu, D.; Srivastava, H.M. Series representations for fractional-calculus operators involving generalised Mittag-Leffler functions. *Commun. Nonlinear Sci. Numer.* **2019**, *67*, 517–527. [CrossRef]
24. Baleanu, D.; Jajarmi, A.; Hajipour, M. On the nonlinear dynamical systems within the generalized fractional derivatives with Mittag-Leffler kernel. *Nonlinear Dyn.* **2018**, *94*, 397–414. . [CrossRef]
25. Qureshi, S.; Yusuf, A.; Shaikh, A.A.; Inc, M.; Baleanu, D. Fractional modeling of blood ethanol concentration system with real data application. *Chaos* **2019**, *29*, 131–143. [CrossRef]
26. Uçar, S.; Uçar, E.; Özdemir, N.; Hammouch, Z. Mathematical analysis and numerical simulation for a smoking model with Atangana-Baleanu derivative. *Chaos Solitons Fractals* **2019**, *118*, 300–306. [CrossRef]
27. Yavuz, M.; Özdemir, N.; Başkonuş, H.M. Solutions of partial differential equations using the fractional operator involving Mittag-Leffler kernel. *Eur. Phys. J. Plus* **2018**, *133*, 1–11. [CrossRef]
28. Fernandez, A.; Husain, I. Modified Mittag-Leffler functions with applications in complex formulae for fractional calculus. *Fractal Fract.* **2020**, *4*, 15. [CrossRef]
29. Xiao, X.; Fu, P.; Hu, G.; Sangiah, A.K.; Zheng, H.; Jiang, Y. SAIDR: A new dynamic model for SMS-based worm propagation in mobile networks. *IEEE Access* **2017**, *5*, 9935–9943. [CrossRef]
30. Baleanu, D.; Fernandez, A. On some new properties of fractional derivatives with Mittag-Leffler kernel. *Commun. Nonlinear Sci. Numer. Simul.* **2018**, *59*, 444–462. [CrossRef]
31. Gómez-Aguilar, J.F.; Rosales-García, J.J.; Bernal-Alvarado, J.J. Fractional mechanical oscillators. *Rev. Mex. FíSica* **2012**, *58*, 348–352.
32. Qing, Y.; Rhoades, B.E. T-stability of Picard iteration in metric spaces. *Fixed Point Theory Appl.* **2008**, *2008*. [CrossRef]
33. Mekkaoui, T.; Atangana, A. New numerical approximation of fractional derivative with non-local and non-singular kernel: Application to chaotic models. *Eur. Phys. J. Plus* **2017**, *132*, 1–16.

 *fractal and fractional*

Article

On a Five-Parameter Mittag-Leffler Function and the Corresponding Bivariate Fractional Operators

Mehmet Ali Özarslan * and Arran Fernandez

Department of Mathematics, Faculty of Arts and Sciences, Eastern Mediterranean University, via Mersin-10, Famagusta 99628, Northern Cyprus, Turkey; arran.fernandez@emu.edu.tr
* Correspondence: mehmetali.ozarslan@emu.edu.tr

Abstract: Several extensions of the classical Mittag-Leffler function, including multi-parameter and multivariate versions, have been used to define fractional integral and derivative operators. In this paper, we consider a function of one variable with five parameters, a special case of the Fox–Wright function. It turns out that the most natural way to define a fractional integral based on this function requires considering it as a function of two variables. This gives rise to a model of bivariate fractional calculus, which is useful in understanding fractional differential equations involving mixed partial derivatives.

Keywords: Mittag-Leffler functions; fractional integrals; fractional derivatives; Abel equations; Laplace transforms; mixed partial derivatives

MSC: 33E12; 26A33; 34A08

Citation: Özarslan, M.A.; Fernandez, A. On a Five-Parameter Mittag-Leffler Function and the Corresponding Bivariate Fractional Operators. *Fractal Fract.* **2021**, *5*, 45. https://doi.org/10.3390/fractalfract5020045

Academic Editor: Carlo Cattani

Received: 29 March 2021
Accepted: 12 May 2021
Published: 14 May 2021

Publisher's Note: MDPI stays neutral with regard to jurisdictional claims in published maps and institutional affiliations.

Copyright: © 2021 by the authors. Licensee MDPI, Basel, Switzerland. This article is an open access article distributed under the terms and conditions of the Creative Commons Attribution (CC BY) license (https://creativecommons.org/licenses/by/4.0/).

1. Introduction

The original Mittag-Leffler function $E_\alpha(z)$, applied to one variable according to one parameter, was first defined and studied by Gösta Mittag-Leffler in the 1900s. In the hundred years since then, many variants and extensions of this function have been defined, including functions of more than one variable and functions with arbitrarily many parameters [1,2].

One of the major motivations for studying Mittag-Leffler functions is their relationship with fractional calculus [2–5], a field of mathematics which has become very popular due to its many applications in various areas of science [6–8]. Mittag-Leffler functions emerge naturally as the solution to some elementary fractional differential equations, and their eigenfunction properties have led them sometimes to be called "fractional exponential functions" [9].

Among the many extensions of the Mittag-Leffler function to more variables and parameters, we mention just a few which are of particular importance or interest in motivating the present work.

- A Mittag-Leffler function of one variable with three parameters was defined by Prabhakar [10] to solve a certain singular integral equation. Its use as an integral kernel gave rise to a model of fractional calculus which has a semigroup property and which is already broad enough to include many other named fractional-calculus operators [11–13], although it is itself a special case of the general Fox–Wright function.
- A Mittag-Leffler function of n variables with $n+1$ parameters was used by Luchko et al. [14,15] to solve multi-term linear fractional differential equations involving multiple independent fractional orders. Note that this is independent from the Mittag-Leffler function of n variables with $2n+1$ parameters which was defined by Saxena et al. [16] and which gives rise to a model of fractional calculus with a semigroup property [17]: neither of these general functions is a special case of the other.

- A Mittag-Leffler function of two variables with four parameters and a Mittag-Leffler function of three variables with five parameters were recently defined [18,19] and used to solve multi-order systems of fractional differential equations [20,21]. These are closely related to the general function considered by Luchko et al., but they are not special cases of it. Used as kernels, they also give rise to new models of fractional calculus with a semigroup property [18,21]. For other types of bivariate Mittag-Leffler function that have been defined in the literature, we refer to the papers [22–24].

All of these Mittag-Leffler type functions have been connected in some way to fractional calculus and fractional differential equations. Indeed, the strong connection between special functions and fractional calculus has been known for decades and continues to be written about today [25–27], while multi-parameter and multi-variable generalisations are being studied both for special functions and for the operators of fractional calculus [17].

Fractional partial differential equations have been an interesting and challenging topic of study [3], with various methods able to be extended (with modifications) from classical partial differential equations in order to solve them, such as the unified transform method [28], distribution theory [29,30], and weak solutions [31,32]. Much attention has been paid to partial differential equations involving the fractional Laplacian operator [33,34], but less attention has gone to differential equations involving mixed partial fractional derivatives. They are mentioned in the classical textbooks [3] (§6.1.1) and [6] (§24.2), and in a few papers (e.g., [35–37]), but in general they have attracted little notice in the research literature on fractional partial differential equations.

In the work below, we study a specific type of Mittag-Leffler function, initially a function of one variable with five parameters, and define a double fractional integral operator by converting this function to a bivariate version. In this way, it is possible to involve double integrals and derivatives, with respect to two variables, while preserving the relatively simple structure of a single power series defining the Mittag-Leffler function. These operators are therefore useful in the study of fractional partial differential equations, including those involving mixed partial fractional derivatives.

As an initial motivation, let us consider the following bivariate Abel equation of the second kind:

$$u(t,s) + \frac{\lambda}{\Gamma(\alpha_1)\Gamma(\alpha_2)} \int_0^t \int_0^s \frac{u(\xi,\eta)}{(t-\tau)^{1-\alpha_1}(s-\eta)^{1-\alpha_2}} \, d\tau \, d\eta = f(t,s). \qquad (1)$$

One can reformulate this equation in terms of the standard Riemann–Liouville fractional integral operators as:

$$\left(1 + \lambda I_t^{\alpha_1} I_s^{\alpha_2}\right) u(t,s) = f(t,s).$$

Formally, by means of symbolic operational calculus without regard for rigour, this can be solved as follows:

$$u(t,s) = \left(1 + \lambda I_t^{\alpha_1} I_s^{\alpha_2}\right)^{-1} f(t,s) = \left(\sum_{n=0}^{\infty} (-\lambda)^n I_t^{\alpha_1 n} I_s^{\alpha_2 n}\right) f(t,s).$$

This last formal result can be written in terms of the following Mittag-Leffler type function:

$$E_{\alpha_1,\alpha_2}(x) = \sum_{n=0}^{\infty} \frac{x^n}{\Gamma(\alpha_1 n + 1)\Gamma(\alpha_2 n + 1)}, \qquad (2)$$

namely, in the following way, making use of the double (two-dimensional) Laplace convolution operator denoted here as "∗":

$$u(t,s) = E_{\alpha_1,\alpha_2}(-\lambda t^{\alpha_1} s^{\alpha_2}) * f(t,s)$$
$$= \int_0^t \int_0^s \left(\sum_{n=0}^{\infty} \frac{\lambda^n (t-\tau)^{\alpha_1 n} (s-\eta)^{\alpha_2 n}}{\Gamma(\alpha_1 n + 1)\Gamma(\alpha_2 n + 1)} \right) f(\tau,\eta) \, d\tau \, d\eta$$
$$= \left(\sum_{n=0}^{\infty} (-\lambda)^n I_t^{\alpha_1 n} I_s^{\alpha_2 n} \right) f(t,s).$$

Therefore, it makes sense to define an integral operator $\mathcal{E}_{\alpha_1,\alpha_2}$ acting on functions of two variables as follows:

$$(\mathcal{E}_{\alpha_1,\alpha_2} f)(t,s) = \int_0^t \int_0^s \left(\sum_{n=0}^{\infty} \frac{\lambda^n (t-\tau)^{\alpha_1 n} (s-\eta)^{\alpha_2 n}}{\Gamma(\alpha_1 n + 1)\Gamma(\alpha_2 n + 1)} \right) f(\tau,\eta) \, d\tau \, d\eta. \quad (3)$$

This operator emerges naturally from consideration of bivariate Abel equations, but it has the drawback of lacking a semigroup property, or index law, in any of the parameters: if we take a composition of this operator with itself, we do *not* find another operator in the same form for different values of the parameters.

It should be noted at this point that some useful and important operators, which have been criticised for lacking a semigroup property, can be embedded into a larger class of fractional-calculus operators which has an index law and which therefore contains both the original operators and their compositions. For example, Prabhakar fractional calculus forms a class of operators which contains various useful operators that lack semigroup properties in themselves, their compositions being *different* elements of the Prabhakar class [11]. We can apply the same way of thinking here, extending the basic two-parameter Mittag-Leffler function (2) and the associated integral operator to more general versions where a semigroup property can be found.

In this case, it is sufficient to add three extra parameters in the definition in order to obtain a generalised operator which has a semigroup property. We consider, throughout this paper, the following five-parameter Mittag-Leffler function of one variable:

$$E_{\alpha_1,\alpha_2;\beta_1,\beta_2}^{\gamma}(z) = \sum_{k=0}^{\infty} \frac{(\gamma)_k}{\Gamma(\alpha_1 k + \beta_1)\Gamma(\alpha_2 k + \beta_2)} \cdot \frac{z^k}{k!}, \quad (4)$$

where $\alpha_1, \alpha_2, \beta_1, \beta_2, \gamma$ are complex constants (with some constraints to be determined later) and $(\gamma)_k$ is the Pochhammer symbol defined by

$$(\gamma)_0 = 1, \quad (\gamma)_k = \gamma(\gamma+1)\cdots(\gamma+k-1), \quad k = 1,2,\cdots.$$

This paper is devoted to a detailed study of the five-parameter Mittag-Leffler function (4), the associated bivariate fractional integral operators, related concepts such as the corresponding fractional derivative operators, and special cases of particular interest.

Specifically, the structure of the paper is as follows. We discuss the five-parameter Mittag-Leffler function and its properties in Section 2, then pass to bivariate fractional calculus in Section 3, firstly integral operators and then derivative operators. In Section 4, we investigate the case where the operators are non-singular and expressible as finite sums of Riemann–Liouville integrals. Section 5 is devoted to discussion of applications and potential future work.

2. The Five-Parameter Mittag-Leffler Function

Definition 1. *Let $\alpha_1, \alpha_2, \beta_1, \beta_2, \gamma \in \mathbb{C}$ be five parameters satisfying $\operatorname{Re}(\alpha_1 + \alpha_2) > 0$. The five-parameter Mittag-Leffler function applied to a single variable z is defined by the following power series:*

$$E^{\gamma}_{\alpha_1,\alpha_2;\beta_1,\beta_2}(z) = \sum_{k=0}^{\infty} \frac{(\gamma)_k}{\Gamma(\alpha_1 k + \beta_1)\Gamma(\alpha_2 k + \beta_2)} \cdot \frac{z^k}{k!}, \qquad z \in \mathbb{C}.$$

This can be seen as a special case of the Fox–Wright function, which is defined [3] (§1.11) by

$$_p\Psi_q(z) = {_p\Psi_q}((a_i, \alpha_i)_{1,p}; (b_i, \beta_i)_{1,q}; z) = \sum_{k=0}^{\infty} \frac{\prod_{i=1}^{p} \Gamma(a_i + \alpha_i k)}{\prod_{j=1}^{q} \Gamma(b_j + \beta_j k)} \cdot \frac{z^k}{k!}, \qquad z \in \mathbb{C},$$

where, in [3], the parameters are given as $a_i, b_j \in \mathbb{C}$, $\alpha_i, \beta_j \in \mathbb{R}$ satisfying $\alpha_i \neq \beta_j$ for $i = 1, \cdots, p; j = 1, \cdots, q$, and it is proved that this power series is absolutely convergent for all $z \in \mathbb{C}$ in the case $\Delta := \sum_{j=1}^{q} \beta_j - \sum_{i=1}^{p} \alpha_i > -1$.

The five-parameter Mittag-Leffler function can be written in terms of the Fox–Wright function as follows:

$$E^{\gamma}_{\alpha_1,\alpha_2;\beta_1,\beta_2}(z) = \frac{1}{\Gamma(\gamma)} {_1\Psi_2}((\gamma, 1); (\beta_1, \alpha_1), (\beta_2, \alpha_2); z).$$

As an immediate consequence of the convergence result shown in [3] (Theorem 1.5) and [38], we conclude that, if $\alpha_1, \alpha_2 \in \mathbb{R}$ and $\alpha_1 + \alpha_2 > 0$, then the power series defining $E^{\gamma}_{\alpha_1,\alpha_2;\beta_1,\beta_2}(z)$ is locally uniformly convergent and therefore it is an entire function.

In fact, there is no need to assume that any of the parameters are real. This assumption is imposed in [3,38] to simplify the calculations, but the same convergence result can be proved for complex values of the parameters by using Stirling's formula and the ratio test, similarly to the second author's work in [39]. The condition then to be imposed on the parameters is $\operatorname{Re}(\alpha_1 + \alpha_2) > 0$, as stated in Definition 1.

It is worth noting that a four-parameter special case of the general Fox–Wright function was recently given particular consideration by Luchko [40], namely the following function:

$$W_{(\alpha_1,\beta_1),(\alpha_2,\beta_2)}(z) = \sum_{k=0}^{\infty} \frac{z^k}{\Gamma(\alpha_1 k + \beta_1)\Gamma(\alpha_2 k + \beta_2)}, \qquad (5)$$

which is the case $\gamma = 1$ of our five-parameter Mittag-Leffler function. It is also a special case of the vector-index Mittag-Leffler function studied by Al-Bassam and Luchko [41], which is defined as follows:

$$E_{(\alpha_1,\beta_1),\cdots,(\alpha_n,\beta_n)}(z) = \sum_{k=0}^{\infty} \frac{z^k}{\Gamma(\alpha_1 k + \beta_1)\cdots\Gamma(\alpha_n k + \beta_n)}, \qquad (6)$$

for any $n \in \mathbb{N}$. The case $n = 1$ gives the usual two-parameter Mittag-Leffler function, while the case $n = 2$ gives the four-parameter Wright function (5) considered by Luchko. The function (6) was further extended [42] to include a numerator $(\gamma)_{\kappa n}$ in terms of two further parameters.

Our function is also a special case of the general Fox–Wright function, but it is different from the special cases (5) and (6) considered previously, because of the Pochhammer symbol appearing in the numerator. This extra parameter and Pochhammer symbol is important because, as we show below, it gives rise to a semigroup property for the resulting fractional integral operators – a property which is lacking for functions such as (5) and (6) which were considered before, and even for functions such as that in [12] which contain a generalised Pochhammer symbol. The semigroup property will arise from the idea of turning this univariate function into a bivariate integral operator, a notion which we justify below from consideration of the gamma functions on the denominator.

Using the Mellin–Barnes integral representation of the Fox–Wright function $_p\Psi_q(z)$, proved in [3], we have for $\alpha_1 > 0, \alpha_2 > 0$, and $\gamma, \beta_1, \beta_2 \in \mathbb{C}, z \neq 0$ that

$$E^{\gamma}_{\alpha_1,\alpha_2;\beta_1,\beta_2}(z) = \frac{1}{2\pi i \Gamma(\gamma)} \int_L \frac{\Gamma(s)\Gamma(\gamma - s)}{\Gamma(\beta_1 - \alpha_1 s)\Gamma(\beta_2 - \alpha_2 s)} (-z)^{-s} ds,$$

where the integration path L is a Bromwich contour starting at a point $C - i\infty$ and terminating at a point $C + i\infty$ and separating the poles $-m$ ($m = 0, 1, 2, \cdots$) of the gamma function $\Gamma(s)$ to the left and the poles $\gamma + l$ ($l = 0, 1, 2, \cdots$) to the right with the assumption $-m \neq \gamma + l$ for $l, m = 0, 1, 2, \cdots$.

Another complex integral representation is given by the following Theorem.

Theorem 1. Let $\alpha_1, \alpha_2, \beta_1, \beta_2 \in \mathbb{C}$ with $\text{Re}(\alpha_1) > 0, \text{Re}(\alpha_2) > 0$, and let $\varepsilon > 0, \frac{\pi}{2} < \phi \leq \pi$. Let $H(\varepsilon; \phi) (\varepsilon > 0, 0 < \phi \leq \pi)$ be the contour which is the union of the following three parts oriented according to non-decreasing $\arg \tau$:

1. the ray $\arg \tau = -\phi, |\tau| \geq \varepsilon$;
2. the arc $-\phi \leq \arg \tau \leq \phi, |\tau| = \varepsilon$; and
3. the ray $\arg \tau = \phi, |\tau| \geq \varepsilon$,

Then, the five-parameter Mittag-Leffler function possesses the following complex integral representations:

$$E^{\gamma}_{\alpha_1,\alpha_2;\beta_1,\beta_2}(z) = \frac{1}{2\pi i} \int_{H(\varepsilon;\phi)} \tau^{-\beta_2} e^{\tau} E^{\gamma}_{\alpha_1,\beta_1}\left(\tau^{-\alpha_2} z\right) d\tau \quad (7)$$

$$= \frac{-1}{4\pi^2} \int_{H(\varepsilon;\phi)} \int_{H(\varepsilon;\phi)} \frac{\xi^{-\beta_1} \tau^{-\beta_2} e^{\xi+\tau}}{(1 - z\xi^{-\alpha_1}\tau^{-\alpha_2})^{\gamma}} d\xi\, d\tau. \quad (8)$$

Proof. Using the known representation [43]

$$\frac{1}{\Gamma(z)} = \frac{1}{2\pi i} \int_{H(\varepsilon;\phi)} \tau^{-z} e^{\tau} d\tau, \quad z \in \mathbb{C}, \quad (9)$$

with z replaced by $\alpha_2 k + \beta_2$, and using the fact that the series represents an entire function, which allows the interchange of the series and the integral, gives

$$E^{\gamma}_{\alpha_1,\alpha_2;\beta_1,\beta_2}(z) = \sum_{k=0}^{\infty} \frac{(\gamma)_k}{\Gamma(\alpha_1 k + \beta_1)} \left(\frac{1}{2\pi i} \int_{H(\varepsilon;\phi)} \tau^{-\alpha_2 k - \beta_2} e^{\tau} d\tau\right) \frac{z^k}{k!}$$

$$= \frac{1}{2\pi i} \int_{H(\varepsilon;\phi)} \tau^{-\beta_2} e^{\tau} \sum_{k=0}^{\infty} \frac{(\gamma)_k}{\Gamma(\alpha_1 k + \beta_1)} \frac{(\tau^{-\alpha_2} z)^k}{k!} d\tau,$$

which directly yields (7), the first of the desired formulae, in terms of the three-parameter Mittag-Leffler function of Prabhakar.

On the other hand, using (9) for both of the terms $\frac{1}{\Gamma(\alpha_1 k + \beta_1)}$ and $\frac{1}{\Gamma(\alpha_2 k + \beta_2)}$, we have

$$E^{\gamma}_{\alpha_1,\alpha_2;\beta_1,\beta_2}(z) = \frac{-1}{4\pi^2} \sum_{k=0}^{\infty} (\gamma)_k \left(\int_{H(\varepsilon;\phi)} \xi^{-\alpha_1 k - \beta_1} e^{\xi} d\xi \cdot \int_{H(\varepsilon;\phi)} \tau^{-\alpha_2 k - \beta_2} e^{\tau} d\tau\right) \frac{z^k}{k!}$$

$$= \frac{-1}{4\pi^2} \int_{H(\varepsilon;\phi)} \int_{H(\varepsilon;\phi)} \xi^{-\beta_1} \tau^{-\beta_2} e^{\xi+\tau} \sum_{k=0}^{\infty} (\gamma)_k \frac{(z\xi^{-\alpha_1}\tau^{-\alpha_2})^k}{k!} d\xi\, d\tau,$$

which directly yields (8), the second of the desired formulae. This derivation assumes that $|z\xi^{-\alpha_1}\tau^{-\alpha_2}| < 1$, which is uniformly true on the given contours provided that $\text{Re}(\alpha_1) > 0$,

Re(α_2) > 0, and $|z|$ is sufficiently small. The bound on $|z|$ can be removed, by analytic continuation in z, to yield the same result valid for all z. □

Thus far, we have considered the function (4) as a five-parameter function of a single variable z. As a function of z, this function has several properties, such as complex integral representations, which we prove above. However, when we start to construct connections with the topic of fractional calculus, it is more natural to consider a related bivariate function instead.

To see why, consider that fractional differintegral relations for special functions, such as many found in [44,45], usually make use of gamma functions appearing in power series, in the following way:

$$_aI_x^\lambda\left(\frac{(x-a)^\nu}{\Gamma(\nu+1)}\right) = \frac{(x-a)^{\nu+\lambda}}{\Gamma(\nu+\lambda+1)}, \qquad _aD_x^\lambda\left(\frac{(x-a)^\nu}{\Gamma(\nu+1)}\right) = \frac{(x-a)^{\nu-\lambda}}{\Gamma(\nu-\lambda+1)},$$

with appropriate choices of ν in order to use such identities for every term of a power series expansion. For many univariate functions defined by power series, the above relations give rise to interesting identities between special functions. In our case, however, the power series (4) has two different gamma functions in the denominator, each of them involving k times a different parameter, α_1 or α_2. Therefore, to take full advantage of the function's symmetry in α_1 and α_2, we should use two different variables: one raised to the power of α_1, the other raised to the power of α_2, and then both of them further raised to the power of k.

Motivated by this discussion, we now begin to study, instead of the univariate function $E^\gamma_{\alpha_1,\alpha_2;\beta_1,\beta_2}(z)$, the closely related bivariate function $E^\gamma_{\alpha_1,\alpha_2;\beta_1,\beta_2}(x^{\alpha_1}y^{\alpha_2})$. A similar idea, substituting products of fractional powers of new variables instead of old variables, was used in a 2017 paper of the first author [23], but in that case it was used to convert a bivariate function to another bivariate function of different variables. Here, we use it to convert a univariate function to a bivariate function. This seems at first sight as an unnecessary complication, but we see that it makes many things more natural and smooth in the studies related to the five-parameter Mittag-Leffler function.

Definition 2 ([3]). *The double fractional integrals of a bivariate function $f(x,t)$ are defined in the natural way by combining fractional integrals with respect to x and t, namely as follows for $\lambda, \mu \in \mathbb{C}$ with $\text{Re}(\lambda) > 0, \text{Re}(\mu) > 0$ and for $x > a, t > b$.*

$$_bI_t^\lambda \, _aI_x^\mu f(x,t) = \frac{1}{\Gamma(\mu)\Gamma(\lambda)}\int_b^t \int_a^x (t-\tau)^{\lambda-1}(x-\xi)^{\mu-1}f(\xi,\tau)\,d\xi\,d\tau.$$

Similarly, the partial fractional derivatives of a bivariate function $f(x,t)$ are defined as follows, for $\lambda, \mu \in \mathbb{C}$ with $\text{Re}(\lambda) > 0, \text{Re}(\mu) > 0$ and for $n = \lfloor \text{Re}(\mu) \rfloor + 1$, $m = \lfloor \text{Re}(\lambda) \rfloor + 1$ and $x > a, t > b$:

$$_bD_t^\beta \, _aD_x^\alpha f(x,t) = \frac{\partial^{m+n}}{\partial t^m \partial x^n} \, _bI_t^{n-\beta} \, _aI_x^{m-\alpha} f(x,t).$$

Lemma 1. *Let $\alpha_1, \alpha_2, \beta_1, \beta_2, \gamma, \omega \in \mathbb{C}$ with $\text{Re}(\alpha_1), \text{Re}(\alpha_2), \text{Re}(\beta_1), \text{Re}(\beta_2) > 0$. Then, for any $\lambda, \mu \in \mathbb{C}$ with $\text{Re}(\lambda) > 0, \text{Re}(\mu) > 0$, we have*

$$_aI_x^\lambda \, _cI_y^\mu \left[(x-a)^{\beta_1-1}(y-c)^{\beta_2-1}E^\gamma_{\alpha_1,\alpha_2;\beta_1,\beta_2}(\omega(x-a)^{\alpha_1}(y-c)^{\alpha_2})\right]$$
$$= (x-a)^{\lambda+\beta_1-1}(y-c)^{\mu+\beta_2-1}E^\gamma_{\alpha_1,\alpha_2;\lambda+\beta_1,\mu+\beta_2}(\omega(x-a)^{\alpha_1}(y-c)^{\alpha_2}).$$

Proof. We know from above that the infinite series defining the five-parameter Mittag-Leffler function is locally uniformly convergent, so we have the right to interchange the order of this series with fractional integral operators. Since all the exponents of $(x-a)$ and $(y-c)$ are greater than -1, we have:

$$_aI_x^\lambda {}_cI_y^\mu \left[(x-a)^{\beta_1-1}(y-c)^{\beta_2-1} E_{\alpha_1,\alpha_2;\beta_1,\beta_2}^\gamma (\omega(x-a)^{\alpha_1}(y-c)^{\alpha_2}) \right]$$

$$= \sum_{k=0}^\infty \frac{(\gamma)_k}{\Gamma(\alpha_1 k + \beta_1)\Gamma(\alpha_2 k + \beta_2)} \cdot \frac{\omega^k}{k!} \left[{}_aI_x^\lambda (x-a)^{\beta_1+\alpha_1 k-1} \right] \left[{}_cI_y^\mu (y-c)^{\beta_2+\alpha_2 k-1} \right]$$

$$= \sum_{k=0}^\infty \frac{(\gamma)_k}{\Gamma(\alpha_1 k + \beta_1)\Gamma(\alpha_2 k + \beta_2)} \cdot \frac{\omega^k}{k!}$$
$$\times \left[\frac{\Gamma(\beta_1 + \alpha_1 k)}{\Gamma(\lambda + \beta_1 + \alpha_1 k)} (x-a)^{\lambda+\beta_1+\alpha_1 k-1} \right] \left[\frac{\Gamma(\beta_2 + \alpha_2 k)}{\Gamma(\mu + \beta_2 + \alpha_2 k)} (y-c)^{\mu+\beta_2+\alpha_2 k-1} \right]$$

$$= (x-a)^{\lambda+\beta_1-1}(y-c)^{\mu+\beta_2-1} E_{\alpha_1,\alpha_2;\lambda+\beta_1,\mu+\beta_2}^\gamma (\omega(x-a)^{\alpha_1}(y-c)^{\alpha_2}).$$

Note how the two gamma functions in the denominator of the series mesh together precisely with the gamma-function quotients arising from the two fractional integrals, in order to achieve the desired result. □

Lemma 2. *Let $\alpha_1, \alpha_2, \beta_1, \beta_2, \gamma, \omega \in \mathbb{C}$ with $\mathrm{Re}(\alpha_1), \mathrm{Re}(\alpha_2), \mathrm{Re}(\beta_1), \mathrm{Re}(\beta_2) > 0$. Then, for any $\lambda, \mu \in \mathbb{C}$ with $\mathrm{Re}(\lambda) \geq 0, \mathrm{Re}(\mu) \geq 0$, we have*

$$_aD_x^\lambda {}_cD_y^\mu \left[(x-a)^{\beta_1-1}(y-c)^{\beta_2-1} E_{\alpha_1,\alpha_2;\beta_1,\beta_2}^\gamma (\omega(x-a)^{\alpha_1}(y-c)^{\alpha_2}) \right]$$
$$= (x-a)^{\beta_1-\lambda-1}(y-c)^{\beta_2-\mu-1} E_{\alpha_1,\alpha_2;\beta_1-\lambda,\beta_2-\mu}^\gamma (\omega(x-a)^{\alpha_1}(y-c)^{\alpha_2}).$$

Proof. This can be deduced from Lemma 1 by analytic continuation in the variables λ and μ, using the analyticity properties of fractional differintegrals. Alternatively, it can be proved from the series expansion of the function, following similar lines as the proof of Lemma 1. □

Lemma 3. *Setting all parameters to 1 in the bivariate form of the five-parameter Mittag-Leffler function, we can recover a case of the modified Bessel function with parameter 0:*

$$E_{1,1;1,1}^1(xy) = I_0(2\sqrt{xy}) = \sum_{k=0}^\infty \frac{(xy)^k}{(k!)^2}.$$

Proof. This follows immediately from the series definition of the function. □

3. Bivariate Operators with Five-Parameter Mittag-Leffler Kernels

In this section, we move on from functions to operators. Having established the five-parameter Mittag-Leffler function and some of its properties, we now wish to define a fractional integral operator using this function as a kernel, following in the footsteps of other papers [10,18,46] which defined new models of fractional calculus by using various types of Mittag-Leffler functions as kernels.

The five-parameter Mittag-Leffler function (4) is defined by a single power series in terms of a single variable z. When we use it to define a fractional integral operator, however, we must transform it to a bivariate function, again using a single summation, but this time in terms of two independent variables x, y. Correspondingly, we use a double integral and create an operator to be applied to bivariate functions. This is the only natural way to define a model of fractional calculus using the five-parameter Mittag-Leffler function, because the two gamma functions in its denominator will give rise to a double fractional integral in each summand when we wish to write a series representation for the new operator: we need two separate powers in the integrand, one corresponding to each of the gamma functions from the denominator.

Definition 3. *The fractional integral operator based on the five-parameter Mittag-Leffler function as a kernel is given by*

$$_{a,b}\mathcal{I}^{\gamma;\lambda}_{\alpha_1,\alpha_2;\beta_1,\beta_2}(f)(x,y) := \int_a^x \int_b^y f(t,s)(x-t)^{\beta_1-1}(y-s)^{\beta_2-1} E^{\gamma}_{\alpha_1,\alpha_2;\beta_1,\beta_2}(\lambda(x-t)^{\alpha_1}(y-s)^{\alpha_2})\,ds\,dt,$$

where $a,b \in \mathbb{R}$ are fixed with $x > a$, $y > b$ and $\alpha_1, \alpha_2, \beta_1, \beta_2, \gamma, \lambda \in \mathbb{C}$ are parameters such that $\mathrm{Re}(\alpha_1), \mathrm{Re}(\alpha_2), \mathrm{Re}(\beta_1), \mathrm{Re}(\beta_2) > 0$. Note that these restrictions on the parameters are necessary in order to have a properly convergent singular integral for all reasonably well-behaved functions f (more details later on the function space for f).

It is clear that the two-parameter bivariate integral operator defined in (3) above is a special case of this new fractional integral operator: $(\mathcal{E}_{\alpha_1,\alpha_2} f)(x,y) = {}_{0,0}\mathcal{I}^{1;\lambda}_{\alpha_1,\alpha_2;1,1}(f)(x,y)$.

Theorem 2. *Let $\alpha_1, \alpha_2, \beta_1, \beta_2, \gamma, \lambda \in \mathbb{C}$ with $\mathrm{Re}(\alpha_1), \mathrm{Re}(\alpha_2), \mathrm{Re}(\beta_1), \mathrm{Re}(\beta_2) > 0$. Then, ${}_{a,b}\mathcal{I}^{\gamma;\lambda}_{\alpha_1,\alpha_2;\beta_1,\beta_2}$ is a bounded operator from the space $L_1([a,c] \times [b,d])$ to itself.*

Proof. Using Fubini's theorem, we have

$$\int_a^c \int_b^d \left| {}_{a,b}\mathcal{I}^{\gamma;\lambda}_{\alpha_1,\alpha_2;\beta_1,\beta_2}(f)(x,y) \right| dy\,dx$$
$$\leq \int_a^c \int_a^x \int_b^d \int_b^y \left| f(t,s)(x-t)^{\beta_1-1}(y-s)^{\beta_2-1} E^{\gamma}_{\alpha_1,\alpha_2;\beta_1,\beta_2}(\lambda(x-t)^{\alpha_1}(y-s)^{\alpha_2}) \right| ds\,dy\,dt\,dx$$
$$= \int_a^c \int_b^d |f(t,s)| \left[\int_0^{c-t} \int_0^{d-s} \left| u^{\beta_1-1} w^{\beta_2-1} E^{\gamma}_{\alpha_1,\alpha_2;\beta_1,\beta_2}(\lambda u^{\alpha_1} w^{\alpha_2}) \right| dw\,du \right] ds\,dt.$$

Since the five-parameter Mittag-Leffler function is an entire function when $\mathrm{Re}(\alpha_1 + \alpha_2) > 0$, we have a bound of the form $E^{\gamma}_{\alpha_1,\alpha_2;\beta_1,\beta_2}(\lambda u^{\alpha_1} w^{\beta_2-1}) \leq C$ on the finite domain $[a,c] \times [b,d]$. Therefore,

$$\left\| {}_{a,b}\mathcal{I}^{\gamma;\lambda}_{\alpha_1,\alpha_2;\beta_1,\beta_2}(f) \right\|_1 := \int_a^c \int_b^d \left| {}_{a,b}\mathcal{I}^{\gamma;\lambda}_{\alpha_1,\alpha_2;\beta_1,\beta_2}(f)(x,y) \right| dy\,dx$$
$$\leq C(c-a)^{\beta_1-1}(d-b)^{\beta_2-1} \|f\|_1.$$

Since $a, b, c, d, \beta_1, \beta_2$ are fixed, the result is proved. □

Theorem 3. *Let $\alpha_1, \alpha_2, \beta_1, \beta_2, \gamma, \lambda \in \mathbb{C}$ with $\mathrm{Re}(\alpha_1), \mathrm{Re}(\alpha_2), \mathrm{Re}(\beta_1), \mathrm{Re}(\beta_2) > 0$, and let $f \in L_1([a,c] \times [b,d])$. Then, the operator ${}_{a,b}\mathcal{I}^{\gamma;\lambda}_{\alpha_1,\alpha_2;\beta_1,\beta_2}$ can be represented as an infinite series of double fractional integrals of Riemann–Liouville type:*

$$_{a,b}\mathcal{I}^{\gamma;\lambda}_{\alpha_1,\alpha_2;\beta_1,\beta_2}(f)(x,y) = \sum_{k=0}^{\infty} \frac{(\gamma)_k \lambda^k}{k!} {}_b I_y^{\beta_2+\alpha_2 k} {}_a I_x^{\beta_1+\alpha_1 k}(f)(x,y), \qquad (10)$$

where the right-hand side is locally uniformly convergent for $x, y \in [a,c] \times [b,d]$.

Proof. Since $E^{\gamma}_{\alpha_1,\alpha_2;\beta_1,\beta_2}(u)$ is an entire function defined by a locally uniformly convergent power series, and $f \in L_1([a,c] \times [b,d])$, we have the right to interchange the order of summation and integration, to yield

$$_{a,b}\mathcal{I}^{\gamma;\lambda}_{\alpha_1,\alpha_2;\beta_1,\beta_2}(f)(x,y) = \int_a^x \int_b^y f(t,s)(x-t)^{\beta_1-1}(y-s)^{\beta_2-1} E^{\gamma}_{\alpha_1,\alpha_2;\beta_1,\beta_2}(\lambda(x-t)^{\alpha_1}(y-s)^{\alpha_2})\,dt\,ds$$

$$= \int_a^x \int_b^y (x-t)^{\beta_1-1}(y-s)^{\beta_2-1} \sum_{k=0}^{\infty} \frac{(\gamma)_k \lambda^k (x-t)^{\alpha_1 k}(y-s)^{\alpha_2 k}}{\Gamma(\alpha_1 k+\beta_1)\Gamma(\alpha_2 k+\beta_2) k!} f(t,s)\,dt\,ds$$

$$= \sum_{k=0}^{\infty} \frac{(\gamma)_k \lambda^k}{k!} \cdot \frac{1}{\Gamma(\alpha_1 k+\beta_1)\Gamma(\alpha_2 k+\beta_2)} \int_a^x \int_b^y (x-t)^{\beta_1+\alpha_1 k-1}(y-s)^{\beta_2+\alpha_2 k-1} f(t,s)\,dt\,ds$$

$$= \sum_{k=0}^{\infty} \frac{(\gamma)_k \lambda^k}{k!}\, _b I_y^{\beta_2+\alpha_2 k}\, _a I_x^{\beta_1+\alpha_1 k}(f)(x,y).$$

Whence the result. □

Before the next results, we need to recall the bivariate Laplace transform, or double Laplace transform, which is applied to bivariate functions $f(x,y)$ in the following way:

$$\mathcal{L}_2[f](p,q) = \int_0^\infty \int_0^\infty e^{-(px+qy)} f(x,y)\,dx\,dy = \mathcal{L}_x \mathcal{L}_y[f](p,q), \qquad \mathrm{Re}(p)>0, \mathrm{Re}(q)>0,$$

provided that this integral is convergent (for example, if $f(x,y)$ is exponentially bounded in both variables).

Theorem 4. *Let $\alpha_1, \alpha_2, \beta_1, \beta_2, \gamma, \lambda \in \mathbb{C}$ with $\mathrm{Re}(\alpha_1), \mathrm{Re}(\alpha_2), \mathrm{Re}(\beta_1), \mathrm{Re}(\beta_2) > 0$. If f is a bivariate function of exponential order and integrable over $[0,\infty) \times [0,\infty)$, then we have*

$$\mathcal{L}_2\left[_{a,b}\mathcal{I}^{\gamma;\lambda}_{\alpha_1,\alpha_2;\beta_1,\beta_2}(f)\right](p,q) = \frac{1}{p^{\beta_1} q^{\beta_2}} \left(1 - \lambda p^{-\alpha_1} q^{-\alpha_2}\right)^{-\gamma} \mathcal{L}_2[f](p,q)$$

for $p, q \in \mathbb{C}$ such that these Laplace transforms exist.

Proof. This follows from the series representation of the operator, using the fact that the series is locally uniformly convergent to interchange the Laplace integration and summation:

$$\mathcal{L}_2\left[_{a,b}\mathcal{I}^{\gamma;\lambda}_{\alpha_1,\alpha_2;\beta_1,\beta_2}(f)\right](p,q) = \sum_{k=0}^{\infty} \frac{(\gamma)_k \lambda^k}{k!} \mathcal{L}_2\left[_b I_y^{\beta_2+\alpha_2 k}\, _a I_x^{\beta_1+\alpha_1 k}(f)\right](p,q)$$

$$= \mathcal{L}_2[f](p,q) \sum_{k=0}^{\infty} \frac{(\gamma)_k \lambda^k}{k!} q^{-\beta_2-\alpha_2 k} p^{-\beta_1-\alpha_1 k}$$

$$= \frac{1}{p^{\beta_1} q^{\beta_2}} \mathcal{L}_2[f](p,q) \sum_{k=0}^{\infty} \frac{(\gamma)_k (\lambda p^{-\alpha_1} q^{-\alpha_2})^k}{k!}$$

$$= \frac{1}{p^{\beta_1} q^{\beta_2}} \left(1 - \lambda p^{-\alpha_1} q^{-\alpha_2}\right)^{-\gamma} \mathcal{L}_2[f](p,q),$$

where we have assumed $|\lambda p^{-\alpha_1} q^{-\alpha_2}| < 1$ for convergence of the binomial series, although this condition can be removed by analytic continuation in the variables p and q. □

As an application of the above result, we can use Laplace transforms to quickly learn the result of applying the new fractional operator to the five-parameter Mittag-Leffler function itself, as follows.

Example 1. *Let $\alpha_1, \alpha_2, \beta_1, \beta_2, \varepsilon_1, \varepsilon_2, \gamma, \lambda, \sigma \in \mathbb{C}$ with $\mathrm{Re}(\alpha_i), \mathrm{Re}(\beta_i), \mathrm{Re}(\varepsilon_i) > 0$ for $i = 1, 2$. Then, we have*

$$\mathcal{L}_2\left[{}_{a,b}\mathcal{I}^{\gamma;\lambda}_{\alpha_1,\alpha_2;\beta_1,\beta_2}\left(x^{\varepsilon_1-1}y^{\varepsilon_2-1}E^{\sigma}_{\alpha_1,\alpha_2;\varepsilon_1,\varepsilon_2}(\lambda x^{\alpha_1}y^{\alpha_2})\right)\right](p,q)$$
$$= \frac{1}{p^{\beta_1}q^{\beta_2}}\left(1-\lambda p^{-\alpha_1}q^{-\alpha_2}\right)^{-\gamma}\mathcal{L}_2\left[x^{\varepsilon_1-1}y^{\varepsilon_2-1}E^{\sigma}_{\alpha_1,\alpha_2;\varepsilon_1,\varepsilon_2}(\lambda x^{\alpha_1}y^{\alpha_2})\right](p,q)$$
$$= \frac{1}{p^{\beta_1}q^{\beta_2}}\left(1-\lambda p^{-\alpha_1}q^{-\alpha_2}\right)^{-\gamma}\frac{1}{p^{\varepsilon_1}q^{\varepsilon_2}}\left(1-\lambda p^{-\alpha_1}q^{-\alpha_2}\right)^{-\sigma}$$
$$= \frac{1}{p^{\beta_1+\varepsilon_1}q^{\beta_2+\varepsilon_2}}\left(1-\lambda p^{-\alpha_1}q^{-\alpha_2}\right)^{-\gamma-\sigma}.$$

Taking inverse Laplace transforms on both sides of this equation, we get the following formula showing how the bivariate fractional integral operator with five-parameter Mittag-Leffler function kernel can be applied to this particular type of function:

$${}_{a,b}\mathcal{I}^{\gamma;\lambda}_{\alpha_1,\alpha_2;\beta_1,\beta_2}\left(x^{\varepsilon_1-1}y^{\varepsilon_2-1}E^{\sigma}_{\alpha_1,\alpha_2;\varepsilon_1,\varepsilon_2}(\lambda x^{\alpha_1}y^{\alpha_2})\right) = x^{\beta_1+\varepsilon_1-1}y^{\beta_2+\varepsilon_2-1}E^{\gamma+\sigma}_{\alpha_1,\alpha_2;\beta_1+\varepsilon_1,\beta_2+\varepsilon_2}(\lambda x^{\alpha_1}y^{\alpha_2}).$$

A very important property of the bivariate fractional calculus defined in this paper is that it has a semigroup property in the variables β_1, β_2, γ, as expressed by the following theorem. There are several different ways to prove this result, as in [18], and we mention here two of them.

Theorem 5. *Let $\alpha_1, \alpha_2, \beta_1, \beta_2, \varepsilon_1, \varepsilon_2, \gamma, \sigma, \lambda \in \mathbb{C}$ with $\mathrm{Re}(\alpha_i), \mathrm{Re}(\beta_i), \mathrm{Re}(\varepsilon_i) > 0$ for $i = 1, 2$. Then, for any $f \in L_1([a,c] \times [b,d])$, we have*

$${}_{a,b}\mathcal{I}^{\gamma;\lambda}_{\alpha_1,\alpha_2;\beta_1,\beta_2}\,{}_{a,b}\mathcal{I}^{\sigma;\lambda}_{\alpha_1,\alpha_2;\varepsilon_1,\varepsilon_2}(f) = {}_{a,b}\mathcal{I}^{\gamma+\sigma;\lambda}_{\alpha_1,\alpha_2;\beta_1+\varepsilon_1,\beta_2+\varepsilon_2}(f).$$

Proof. In the case that $a = b = 0$ and f is a function whose Laplace transform exists, we know from Theorem 4 that applying the operator ${}_{a,b}\mathcal{I}^{\gamma;\lambda}_{\alpha_1,\alpha_2;\beta_1,\beta_2}$ corresponds, in the Laplace domain, to multiplication by $p^{-\beta_1}q^{-\beta_2}(1-\lambda p^{-\alpha_1}q^{-\alpha_2})^{-\gamma}$. The latter operation clearly has a semigroup property in the parameters β_1, β_2, γ, since these appear only as exponents. Thus, the desired result is clear in this case.

For the general case, we must proceed by manipulation of infinite series and gamma functions, using Theorem 3:

$${}_{a,b}\mathcal{I}^{\gamma;\lambda}_{\alpha_1,\alpha_2;\beta_1,\beta_2}\,{}_{a,b}\mathcal{I}^{\sigma;\lambda}_{\alpha_1,\alpha_2;\varepsilon_1,\varepsilon_2}(f) = \sum_{k=0}^{\infty}\frac{(\gamma)_k\lambda^k}{k!}\,{}_bI^{\beta_2+\alpha_2 k}_y\,{}_aI^{\beta_1+\alpha_1 k}_x \sum_{m=0}^{\infty}\frac{(\sigma)_m\lambda^m}{m!}\,{}_bI^{\varepsilon_2+\alpha_2 m}_y\,{}_aI^{\varepsilon_1+\alpha_1 m}_x(f)$$
$$= \sum_{k=0}^{\infty}\sum_{m=0}^{\infty}\frac{(\gamma)_k(\sigma)_m\lambda^{k+m}}{k!m!}\,{}_bI^{\beta_2+\varepsilon_2+\alpha_2(k+m)}_y\,{}_aI^{\beta_1+\varepsilon_1+\alpha_1(k+m)}_x(f)$$
$$= \sum_{n=0}^{\infty}\left[\sum_{k+m=n}\frac{(\gamma)_k(\sigma)_m}{k!m!}\right]\lambda^n\,{}_bI^{\beta_2+\varepsilon_2+\alpha_2 n}_y\,{}_aI^{\beta_1+\varepsilon_1+\alpha_1 n}_x(f)$$
$$= \sum_{n=0}^{\infty}\left[\frac{(\gamma+\sigma)_n}{n!}\right]\lambda^n\,{}_bI^{\beta_2+\varepsilon_2+\alpha_2 n}_y\,{}_aI^{\beta_1+\varepsilon_1+\alpha_1 n}_x(f)$$
$$= {}_{a,b}\mathcal{I}^{\gamma+\sigma;\lambda}_{\alpha_1,\alpha_2;\beta_1+\varepsilon_1,\beta_2+\varepsilon_2}(f),$$

where for the part in square brackets we use a finite-sum identity on gamma functions (see [12], Theorem 2.9)). □

The semigroup property helps us to obtain the left inverse of the bivariate integral operator constructed above, which will motivate the definition of a fractional derivative operator based on the five-parameter Mittag-Leffler function.

Theorem 6. Let $\alpha_1, \alpha_2, \beta_1, \beta_2, \gamma, \lambda \in \mathbb{C}$ with $\operatorname{Re}(\alpha_1), \operatorname{Re}(\alpha_2), \operatorname{Re}(\beta_1), \operatorname{Re}(\beta_2) > 0$. For any $\varepsilon_1, \varepsilon_2 \in \mathbb{C}$ with $\operatorname{Re}(\varepsilon_1), \operatorname{Re}(\varepsilon_2) > 0$, the following operator is a left inverse to the bivariate fractional integral operator ${}_{a,b}\mathcal{I}^{\gamma;\lambda}_{\alpha_1,\alpha_2;\beta_1,\beta_2}$ considered above:
$$ {}_aD_x^{\beta_1+\varepsilon_1} {}_bD_y^{\beta_2+\varepsilon_2} \circ {}_{a,b}\mathcal{I}^{-\gamma;\lambda}_{\alpha_1,\alpha_2;\varepsilon_1,\varepsilon_2}. $$

Proof. Using the semigroup property, we have
$$ {}_{a,b}\mathcal{I}^{-\gamma;\lambda}_{\alpha_1,\alpha_2;\varepsilon_1,\varepsilon_2} \circ {}_{a,b}\mathcal{I}^{\gamma;\lambda}_{\alpha_1,\alpha_2;\beta_1,\beta_2} = {}_{a,b}\mathcal{I}^{0;\lambda}_{\alpha_1,\alpha_2;\beta_1+\varepsilon_1,\beta_2+\varepsilon_2} = {}_bI_y^{\beta_2+\varepsilon_2} {}_aI_x^{\beta_1+\varepsilon_1}, $$

using the fact that the Pochhammer symbol $(0)_k$ equals 1 if $k = 0$ and equals 0 for all integer $k \geq 1$. Therefore, applying the double Riemann–Liouville fractional differential operator ${}_aD_x^{\beta_1+\varepsilon_1} {}_bD_y^{\beta_2+\varepsilon_2}$ from the left on both sides of the equation, the right-hand side gives the identity operator:
$$ {}_aD_x^{\beta_1+\varepsilon_1} {}_bD_y^{\beta_2+\varepsilon_2} \circ {}_{a,b}\mathcal{I}^{-\gamma;\lambda}_{\alpha_1,\alpha_2;\varepsilon_1,\varepsilon_2} \circ {}_{a,b}\mathcal{I}^{\gamma;\lambda}_{\alpha_1,\alpha_2;\beta_1,\beta_2} = {}_aD_x^{\beta_1+\varepsilon_1} {}_bD_y^{\beta_2+\varepsilon_2} \circ {}_bI_y^{\beta_2+\varepsilon_2} {}_aI_x^{\beta_1+\varepsilon_1} = \mathcal{I}. $$

Thus, we have found the left inverse operator of ${}_{a,b}\mathcal{I}^{\gamma;\lambda}_{\alpha_1,\alpha_2;\beta_1,\beta_2}$ as required. □

Remark 1. It should be remarked that the left inverse operator is independent of the parameters ε_1 and ε_2. The easiest way to see this is by using the series formula:

$$\begin{aligned} {}_aD_x^{\beta_1+\varepsilon_1} {}_bD_y^{\beta_2+\varepsilon_2} \circ {}_{a,b}\mathcal{I}^{-\gamma;\lambda}_{\alpha_1,\alpha_2;\varepsilon_1,\varepsilon_2}(f)(x,y) &= {}_aD_x^{\beta_1+\varepsilon_1} {}_bD_y^{\beta_2+\varepsilon_2} \sum_{k=0}^{\infty} \frac{(-\gamma)_k \lambda^k}{k!} {}_bI_y^{\varepsilon_2+\alpha_2 k} {}_aI_x^{\varepsilon_1+\alpha_1 k}(f)(x,y) \\ &= \sum_{k=0}^{\infty} \frac{(-\gamma)_k \lambda^k}{k!} {}_aD_x^{\beta_1+\varepsilon_1} {}_bD_y^{\beta_2+\varepsilon_2} {}_bI_y^{\varepsilon_2+\alpha_2 k} {}_aI_x^{\varepsilon_1+\alpha_1 k}(f)(x,y) \\ &= \sum_{k=0}^{\infty} \frac{(-\gamma)_k \lambda^k}{k!} {}_bI_y^{-\beta_2+\alpha_2 k} {}_aI_x^{-\beta_1+\alpha_1 k}(f)(x,y), \end{aligned} \quad (11)$$

which is independent of ε_1 and ε_2. Here, we make use of the semigroup property for Riemann–Liouville fractional calculus, in the case where the inner operator is a fractional integral, and we also use the convention (valid by analytic continuation in the order of integration) that a Riemann–Liouville fractional integral to negative order means a Riemann–Liouville fractional derivative, ${}_cI_t^{-\nu} = {}_cD_t^{\nu}$.

Since we can choose any values of ε_1 and ε_2 with positive real part and get the same left inverse operator, we opt for the values which give ordinary (non-fractional) derivatives of order $\beta_1 + \varepsilon_1$ and $\beta_2 + \varepsilon_2$. This means choosing, just like in Riemann–Liouville fractional calculus, $\varepsilon_1 = N_1 - \beta_1$ and $\varepsilon_2 = N_2 - \beta_2$, to obtain the following definition.

Definition 4. Let $\alpha_1, \alpha_2, \beta_1, \beta_2, \gamma, \lambda \in \mathbb{C}$ with $\operatorname{Re}(\alpha_1) > 0, \operatorname{Re}(\alpha_2) > 0, \operatorname{Re}(\beta_1) \geq 0, \operatorname{Re}(\beta_2) \geq 0$. Define $N_1, N_2 \in \mathbb{N}$ to be the numbers such that $N_1 - 1 \leq \operatorname{Re}(\beta_1) < N_1$ and $N_2 - 1 \leq \operatorname{Re}(\beta_2) < N_2$. Then, the double Riemann–Liouville-type fractional derivative with five-parameter Mittag-Leffler kernel is defined by
$$ {}_{a,b}\mathcal{D}^{\gamma;\lambda}_{\alpha_1,\alpha_2;\beta_1,\beta_2}(f)(x,y) = \frac{\partial^{N_1+N_2}}{\partial x^{N_1} \partial y^{N_2}} {}_{a,b}\mathcal{I}^{-\gamma;\lambda}_{\alpha_1,\alpha_2;N_1-\beta_1,N_2-\beta_2}(f)(x,y), $$

while the double Caputo-type fractional derivative with five-parameter Mittag-Leffler kernel is defined by
$$ {}_{a,b}^{C}\mathcal{D}^{\gamma;\lambda}_{\alpha_1,\alpha_2;\beta_1,\beta_2}(f)(x,y) = {}_{a,b}\mathcal{I}^{-\gamma;\lambda}_{\alpha_1,\alpha_2;N_1-\beta_1,N_2-\beta_2}\left(\frac{\partial^{N_1+N_2}}{\partial x^{N_1} \partial y^{N_2}} f\right)(x,y). $$

Theorem 7. Let $\alpha_1, \alpha_2, \beta_1, \beta_2, \gamma, \lambda \in \mathbb{C}$ with $\operatorname{Re}(\alpha_1) > 0, \operatorname{Re}(\alpha_2) > 0, \operatorname{Re}(\beta_1) > 0, \operatorname{Re}(\beta_2) > 0$. The fractional integrals and derivatives with five-parameter Mittag-Leffler kernel have the following inversion properties:

$$_{a,b}\mathcal{D}^{\gamma;\lambda}_{\alpha_1,\alpha_2;\beta_1,\beta_2} \, _{a,b}\mathcal{I}^{\gamma;\lambda}_{\alpha_1,\alpha_2;\beta_1,\beta_2}(f)(x,y) = f(x,y),$$

$$_{a,b}\mathcal{I}^{\gamma;\lambda}_{\alpha_1,\alpha_2;\beta_1,\beta_2} \, _{a,b}^{C}\mathcal{D}^{\gamma;\lambda}_{\alpha_1,\alpha_2;\beta_1,\beta_2}(f)(x,y) = f(x,y) - \sum_{n=0}^{N_1-1} \frac{(x-a)^n}{n!} \cdot \frac{\partial^n}{\partial x^n} f(a,y) - \sum_{m=0}^{N_2-1} \frac{(y-b)^m}{m!} \cdot \frac{\partial^m}{\partial y^m} f(x,b)$$

$$+ \sum_{n=0}^{N_1-1} \sum_{m=0}^{N_2-1} \frac{(x-a)^n (y-b)^m}{n!m!} \cdot \frac{\partial^{n+m}}{\partial x^n \partial y^m} f(a,b).$$

Proof. The first identity is proved during the construction of the left inverse. For the second one, we again use the semigroup property given by Theorem 5:

$$_{a,b}\mathcal{I}^{\gamma;\lambda}_{\alpha_1,\alpha_2;\beta_1,\beta_2} \, _{a,b}^{C}\mathcal{D}^{\gamma;\lambda}_{\alpha_1,\alpha_2;\beta_1,\beta_2}(f)(x,y) = \, _{a,b}\mathcal{I}^{\gamma;\lambda}_{\alpha_1,\alpha_2;\beta_1,\beta_2} \, _{a,b}\mathcal{I}^{-\gamma;\lambda}_{\alpha_1,\alpha_2;N_1-\beta_1,N_2-\beta_2} \left(\frac{\partial^{N_1+N_2}}{\partial x^{N_1} \partial y^{N_2}} f \right)(x,y)$$

$$= \, _{a,b}\mathcal{I}^{0;\lambda}_{\alpha_1,\alpha_2;N_1,N_2} \left(\frac{\partial^{N_1+N_2}}{\partial x^{N_1} \partial y^{N_2}} f \right)(x,y)$$

$$= \, _{a}I^{N_1}_x \, _{b}I^{N_2}_y \left(\frac{\partial^{N_1}}{\partial x^{N_1}} \frac{\partial^{N_2}}{\partial y^{N_2}} f \right)(x,y)$$

$$= f(x,y) - \sum_{n=0}^{N_1-1} \frac{(x-a)^n}{n!} \cdot \frac{\partial^n}{\partial x^n} f(a,y) - \sum_{m=0}^{N_2-1} \frac{(y-b)^m}{m!} \cdot \frac{\partial^m}{\partial y^m} f(x,b) \quad (12)$$

$$+ \sum_{n=0}^{N_1-1} \sum_{m=0}^{N_2-1} \frac{(x-a)^n (y-b)^m}{n!m!} \cdot \frac{\partial^{n+m}}{\partial x^n \partial y^m} f(a,b),$$

by using twice the formula for the nth integral of the nth derivative. □

Proposition 1. Let $\alpha_1, \alpha_2, \beta_1, \beta_2, \gamma, \lambda \in \mathbb{C}$ with $\operatorname{Re}(\alpha_1) > 0, \operatorname{Re}(\alpha_2) > 0, \operatorname{Re}(\beta_1) \geq 0, \operatorname{Re}(\beta_2) \geq 0$. Then, the fractional derivatives with five-parameter Mittag-Leffler kernel, of both Riemann–Liouville type and Caputo type, can be expressed by series formulae as follows:

$$_{a,b}\mathcal{D}^{\gamma;\lambda}_{\alpha_1,\alpha_2;\beta_1,\beta_2}(f)(x,y) = \sum_{k=0}^{\infty} \frac{(-\gamma)_k \lambda^k}{k!} \, _{b}I^{-\beta_2+\alpha_2 k}_y \, _{a}I^{-\beta_1+\alpha_1 k}_x (f)(x,y); \quad (13)$$

$$_{a,b}^{C}\mathcal{D}^{\gamma;\lambda}_{\alpha_1,\alpha_2;\beta_1,\beta_2}(f)(x,y) = \sum_{k=0}^{\infty} \frac{(-\gamma)_k \lambda^k}{k!} \, _{b}I^{N_2-\beta_2+\alpha_2 k}_y \, _{a}I^{N_1-\beta_1+\alpha_1 k}_x \left(\frac{\partial^{N_1+N_2}}{\partial x^{N_1} \partial y^{N_2}} f \right)(x,y). \quad (14)$$

Proof. This follows immediately from Theorem 3 and the series formula (11). □

Remark 2. *From comparing the series formula (10) for fractional integrals with the series formula (13) for Riemann–Liouville-type fractional derivatives, it is now clear that the latter is an analytic continuation of the former, with β_1, β_2, γ replaced by $-\beta_1, -\beta_2, -\gamma$. To see this, we make use of the fact, alluded to in Remark 1 above, that the Riemann–Liouville fractional derivative $_{c}D^{-\nu}_t f(t) = \, _{c}I^{\nu}_t f(t)$, $\operatorname{Re}(\nu) \leq 0$, is the analytic continuation in ν of the fractional integral $_{c}I^{\nu}_t f(t)$, $\operatorname{Re}(\nu) > 0$. Therefore, each term $_{b}I^{-\beta_2+\alpha_2 k}_y \, _{a}I^{-\beta_1+\alpha_1 k}_x (f)(x,y)$ appearing in (13) is exactly the same, under the analytic continuation of Riemann–Liouville differintegrals, as the term $_{b}I^{\beta_2+\alpha_2 k}_y \, _{a}I^{\beta_1+\alpha_1 k}_x (f)(x,y)$ appearing in (10), after replacing β_1, β_2, γ by $-\beta_1, -\beta_2, -\gamma$ as stated.*

Thus, we can adopt the notational convention that

$$_{a,b}\mathcal{I}^{-\gamma;\lambda}_{\alpha_1,\alpha_2;-\beta_1,-\beta_2}(f)(x,y) = \, _{a,b}\mathcal{D}^{\gamma;\lambda}_{\alpha_1,\alpha_2;\beta_1,\beta_2}(f)(x,y), \quad (15)$$

and use this to extend the meaning of both $_{a,b}\mathcal{I}^{\gamma;\lambda}_{\alpha_1,\alpha_2;\beta_1,\beta_2}(f)(x,y)$ and $_{a,b}\mathcal{D}^{\gamma;\lambda}_{\alpha_1,\alpha_2;\beta_1,\beta_2}(f)(x,y)$ to the entire complex plane for the parameters β_1, β_2, without any need to impose conditions on the signs of their real parts. This identity is achieved by analytic continuation in the complex variables β_1, β_2 from one half-plane to the other.

Proposition 2. *Let $\alpha_1, \alpha_2, \beta_1, \beta_2, \gamma, \lambda \in \mathbb{C}$ with $\operatorname{Re}(\alpha_1) > 0, \operatorname{Re}(\alpha_2) > 0, \operatorname{Re}(\beta_1) \geq 0, \operatorname{Re}(\beta_2) \geq 0$. Then, the fractional derivatives of Riemann–Liouville type and Caputo type, with five-parameter Mittag-Leffler kernel, are related to each other as follows:*

$$\begin{aligned}
_{a,b}^{C}\mathcal{D}^{\gamma;\lambda}_{\alpha_1,\alpha_2;\beta_1,\beta_2}(f)(x,y) &= {}_{a,b}\mathcal{D}^{\gamma;\lambda}_{\alpha_1,\alpha_2;\beta_1,\beta_2}(f)(x,y) \\
&\quad - \sum_{k=0}^{\infty} \frac{(-\gamma)_k \lambda^k}{k!} \sum_{n=0}^{N_1-1} \frac{(x-a)^{n-\beta_1+\alpha_1 k}}{\Gamma(n-\beta_1+\alpha_1 k+1)} \cdot {}_b I_y^{-\beta_2+\alpha_2 k} \frac{\partial^n}{\partial x^n} f(a,y) \\
&\quad - \sum_{k=0}^{\infty} \frac{(-\gamma)_k \lambda^k}{k!} \sum_{m=0}^{N_2-1} \frac{(y-b)^{m-\beta_2+\alpha_2 k}}{\Gamma(m-\beta_2+\alpha_2 k+1)} \cdot {}_a I_x^{-\beta_1+\alpha_1 k} \frac{\partial^m}{\partial y^m} f(x,b) \\
&\quad + \sum_{k=0}^{\infty} \frac{(-\gamma)_k \lambda^k}{k!} \sum_{n=0}^{N_1-1} \sum_{m=0}^{N_2-1} \frac{(x-a)^{n-\beta_1+\alpha_1 k}(y-b)^{m-\beta_2+\alpha_2 k}}{\Gamma(n-\beta_1+\alpha_1 k+1)\Gamma(m-\beta_2+\alpha_2 k+1)} \cdot \frac{\partial^{n+m}}{\partial x^n \partial y^m} f(a,b).
\end{aligned}$$

Proof. Starting from the series formulae (13) and (14), we have:

$$\begin{aligned}
_{a,b}^{C}\mathcal{D}^{\gamma;\lambda}_{\alpha_1,\alpha_2;\beta_1,\beta_2}(f)(x,y) &= \sum_{k=0}^{\infty} \frac{(-\gamma)_k \lambda^k}{k!} {}_b I_y^{-\beta_2+\alpha_2 k} {}_a I_x^{-\beta_1+\alpha_1 k} \left({}_b I_y^{N_2} {}_a I_x^{N_1} \frac{\partial^{N_1+N_2}}{\partial x^{N_1} \partial y^{N_2}} f \right)(x,y) \\
&= \sum_{k=0}^{\infty} \frac{(-\gamma)_k \lambda^k}{k!} {}_b I_y^{-\beta_2+\alpha_2 k} {}_a I_x^{-\beta_1+\alpha_1 k} \bigg(f(x,y) - \sum_{n=0}^{N_1-1} \frac{(x-a)^n}{n!} \cdot \frac{\partial^n}{\partial x^n} f(a,y) \\
&\quad - \sum_{m=0}^{N_2-1} \frac{(y-b)^m}{m!} \cdot \frac{\partial^m}{\partial y^m} f(x,b) + \sum_{n=0}^{N_1-1} \sum_{m=0}^{N_2-1} \frac{(x-a)^n(y-b)^m}{n!m!} \cdot \frac{\partial^{n+m}}{\partial x^n \partial y^m} f(a,b) \bigg) \\
&= {}_{a,b}\mathcal{D}^{\gamma;\lambda}_{\alpha_1,\alpha_2;\beta_1,\beta_2}(f)(x,y) - \sum_{k=0}^{\infty} \frac{(-\gamma)_k \lambda^k}{k!} \sum_{n=0}^{N_1-1} \frac{(x-a)^{n-\beta_1+\alpha_1 k}}{\Gamma(n-\beta_1+\alpha_1 k+1)} \cdot {}_b I_y^{-\beta_2+\alpha_2 k} \frac{\partial^n}{\partial x^n} f(a,y) \\
&\quad - \sum_{k=0}^{\infty} \frac{(-\gamma)_k \lambda^k}{k!} \sum_{m=0}^{N_2-1} \frac{(y-b)^{m-\beta_2+\alpha_2 k}}{\Gamma(m-\beta_2+\alpha_2 k+1)} \cdot {}_a I_x^{-\beta_1+\alpha_1 k} \frac{\partial^m}{\partial y^m} f(x,b) \\
&\quad + \sum_{k=0}^{\infty} \frac{(-\gamma)_k \lambda^k}{k!} \sum_{n=0}^{N_1-1} \sum_{m=0}^{N_2-1} \frac{(x-a)^{n-\beta_1+\alpha_1 k}(y-b)^{m-\beta_2+\alpha_2 k}}{\Gamma(n-\beta_1+\alpha_1 k+1)\Gamma(m-\beta_2+\alpha_2 k+1)} \cdot \frac{\partial^{n+m}}{\partial x^n \partial y^m} f(a,b),
\end{aligned}$$

where we use both the formula (12) for the double nth integral of the double nth derivative and also the well-known formulae for Riemann–Liouville fractional differintegrals of power functions. □

4. The Non-Singular Cases of the Operators

Recently, the second author [11] made a detailed study of Prabhakar fractional calculus, separating this class of operators into several subclasses according to their properties. Of particular importance is the consideration of whether an operator is singular or non-singular, and whether the series formula expressing it in terms of Riemann–Liouville integrals is a finite or infinite sum. The same considerations can be applied to other types of fractional calculus, such as the one we are studying here.

In the case of Prabhakar, it was proved [11] (Theorem 4.5) that the most special case is when the operators are non-singular and the sum is finite, and this subclass of

Prabhakar fractional calculus consists precisely of the integer powers of the following (inverse) operators:

$$\mathcal{M}f(t) = f(t) - \omega \,_a I_t^\alpha f(t),$$

$$\mathcal{N}f(t) = \frac{d}{dt}\int_a^t E_\alpha\big(\omega(t-\tau)^\alpha\big) f(\tau)\,d\tau.$$

These operators are obtained [47] by setting $\beta = 0$ (for non-singular operators) and $\rho = \pm 1$ (for finite sums) in the operator $\,_a^P I_t^{\alpha,\beta,\rho,\omega}$ of Prabhakar integration. We may try to do something similar for the bivariate operator of integration with five-parameter Mittag-Leffler kernel.

Let us firstly compare the Prabhakar fractional integral with the bivariate fractional integral considered in this paper. We have

$$\,_a^P I_t^{\alpha,\beta,\rho,\omega} f(t) = \int_a^t f(\tau)(t-\tau)^{\beta-1} E_{\alpha,\beta}^\rho\big(\omega(t-\tau)^\alpha\big)\,d\tau$$

$$= \sum_{k=0}^\infty \frac{(\rho)_k \omega^k}{k!} \,_a I_t^{\beta+\alpha k} f(t),$$

versus

$$_{a,b}\mathcal{I}_{\alpha_1,\alpha_2;\beta_1,\beta_2}^{\gamma;\lambda}(f)(x,y) = \int_a^x \int_b^y f(t,s)(x-t)^{\beta_1-1}(y-s)^{\beta_2-1} E_{\alpha_1,\alpha_2;\beta_1,\beta_2}^\gamma\big(\lambda(x-t)^{\alpha_1}(y-s)^{\alpha_2}\big)\,ds\,dt$$

$$= \sum_{k=0}^\infty \frac{(\gamma)_k \lambda^k}{k!} \,_b I_y^{\beta_2+\alpha_2 k} \,_a I_x^{\beta_1+\alpha_1 k}(f)(x,y).$$

It is clear that our parameters α_1, α_2 correspond to the α of Prabhakar, our β_1, β_2 correspond to the β of Prabhakar, our γ corresponds to the ρ of Prabhakar, and our λ corresponds to the ω of Prabhakar, using all notation as above.

Therefore, for our operator, the process of obtaining a non-singular finite-sum version should involve setting $\beta_1 = \beta_2 = 0$ and $\gamma = \pm 1$. Note that, since Definition 3 for the fractional integral operator requires $\mathrm{Re}(\beta_1) > 0$ and $\mathrm{Re}(\beta_2) > 0$, we must use Definition 4 for the case when $\beta_1 = \beta + 2 = 0$. We find:

$$_{a,b}\mathcal{I}_{\alpha_1,\alpha_2;0,0}^{-1;\lambda}(f)(x,y) = \,_{a,b}\mathcal{D}_{\alpha_1,\alpha_2;0,0}^{1;\lambda}(f)(x,y) = \frac{\partial^2}{\partial x \partial y}\,_{a,b}\mathcal{I}_{\alpha_1,\alpha_2;1,1}^{-1;\lambda}(f)(x,y)$$

$$= \frac{\partial^2}{\partial x \partial y}\int_a^x \int_b^y f(t,s) E_{\alpha_1,\alpha_2;1,1}^{-1}\big(\lambda(x-t)^{\alpha_1}(y-s)^{\alpha_2}\big)\,ds\,dt$$

$$= \frac{\partial^2}{\partial x \partial y}\int_a^x \int_b^y f(t,s)\left[1 - \frac{\lambda(x-t)^{\alpha_1}(y-s)^{\alpha_2}}{\Gamma(\alpha_1+1)\Gamma(\alpha_2+1)}\right]ds\,dt$$

$$= f(x,y) - \lambda\frac{\partial^2}{\partial x \partial y}\,_b I_y^{\alpha_2+1}\,_a I_x^{\alpha_1+1}(f)(x,y)$$

$$= f(x,y) - \lambda\,_b I_y^{\alpha_2}\,_a I_x^{\alpha_1}(f)(x,y). \tag{16}$$

The same formula can also be obtained more directly from the series formula (13):

$$_{a,b}\mathcal{I}_{\alpha_1,\alpha_2;0,0}^{-1;\lambda}(f)(x,y) = \sum_{k=0}^\infty \frac{(-1)_k \lambda^k}{k!}\,_b I_y^{\alpha_2 k}\,_a I_x^{\alpha_1 k}(f)(x,y)$$

$$= f(x,y) - \lambda\,_b I_y^{\alpha_2}\,_a I_x^{\alpha_1}(f)(x,y),$$

where again we use the fact that $(-1)_k$ equals 1 if $k = 0$, -1 if $k = 1$, and 0 for all $k \geq 2$.

The inverse operator is given by setting $\gamma = 1$ instead of $\gamma = -1$, but this one cannot be written in such an elementary way:

$$_{a,b}\mathcal{I}^{1;\lambda}_{\alpha_1,\alpha_2;0,0}(f)(x,y) = {}_{a,b}\mathcal{D}^{-1;\lambda}_{\alpha_1,\alpha_2;0,0}(f)(x,y) = \frac{\partial^2}{\partial x \partial y}{}_{a,b}\mathcal{I}^{1;\lambda}_{\alpha_1,\alpha_2;1,1}(f)(x,y)$$

$$= \frac{\partial^2}{\partial x \partial y}\int_a^x \int_b^y f(t,s) E^1_{\alpha_1,\alpha_2;1,1}(\lambda(x-t)^{\alpha_1}(y-s)^{\alpha_2})\,ds\,dt$$

$$= \frac{\partial^2}{\partial x \partial y}\int_a^x \int_b^y f(t,s)\left[\sum_{k=0}^{\infty}\frac{\lambda^k(x-t)^{\alpha_1 k}(y-s)^{\alpha_2 k}}{\Gamma(\alpha_1 k + 1)\Gamma(\alpha_2 k + 1)}\right]ds\,dt, \quad (17)$$

or equivalently, using the series formula (13),

$$_{a,b}\mathcal{I}^{1;\lambda}_{\alpha_1,\alpha_2;0,0}(f)(x,y) = \sum_{k=0}^{\infty}\lambda^k {}_bI_y^{\alpha_2 k}{}_aI_x^{\alpha_1 k}(f)(x,y). \quad (18)$$

The pair of integro-differential operators given by (16) and (17) is of course reminiscent of the so-called Atangana–Baleanu (AB) integral and derivative [48,49]. The latter are defined similarly: the integral by a linear combination of a function with its Riemann–Liouville fractional integral and the derivative by a derivative of an integral transform involving a Mittag-Leffler function kernel. The new development here is that the operators are now bivariate: we can think of the operators (16) and (17) as forming a two-dimensional Atangana–Baleanu calculus.

We also note that there is now a direct connection with the bivariate Abel equation (1) which we considered at first to motivate this paper. The aforementioned Abel equation was rewritten in a form involving the operator (16), and its solution was constructed in a form involving the inverse operator (17).

It is also important to realise that both operators (16) and (17) involve mixed partial integro-differential operators: the AB-type integral operator (16) is defined using fractional integrals with respect to x and y together, and the AB-type derivative operator (17) is defined using mixed partial derivatives with respect to x and y. Thus, these operators may be useful in the study and understanding of fractional PDEs involving mixed partial derivatives, which we note above are under-appreciated in fractional calculus.

To fully recover a bivariate analogue of the AB operators, we need to choose a value for the parameter λ so that appropriate boundary conditions are realised when the parameters α_1, α_2 go to 0 or 1. This is the reason for the choice of multipliers in the definition of the AB integral and derivative to order α: so that the ordinary first-order integral and derivative are recovered when $\alpha \to 1$ and the original function itself when $\alpha \to 0$. Thus, the following definition is motivated.

Definition 5. *The mixed bivariate AB integral is defined to be*

$$_{a,b}^{AB}I_{x,y}^{\alpha_1,\alpha_2}f(x,y) = \frac{(1-\alpha_1)(1-\alpha_2)}{B(\alpha_1,\alpha_2)}f(x,y) + \frac{\alpha_1\alpha_2}{B(\alpha_1,\alpha_2)}{}_bI_y^{\alpha_2}{}_aI_x^{\alpha_1}(f)(x,y),$$

where $B(\alpha_1,\alpha_2)$ is a normalisation function which is assumed to satisfy $B(0,0) = 1$ and $B(1,1) = 1$.

The mixed bivariate AB derivatives, of Riemann–Liouville and Caputo type, respectively, are defined to be

$$_{a,b}^{ABR}D_{x,y}^{\alpha_1,\alpha_2}f(x,y)$$

$$= \frac{B(\alpha_1,\alpha_2)}{(1-\alpha_1)(1-\alpha_2)}\cdot\frac{\partial^2}{\partial x\partial y}\int_a^x\int_b^y f(t,s)E^1_{\alpha_1,\alpha_2;1,1}\left(\frac{-\alpha_1\alpha_2}{(1-\alpha_1)(1-\alpha_2)}(x-t)^{\alpha_1}(y-s)^{\alpha_2}\right)ds\,dt$$

and

$$^{ABC}_{a,b}D^{\alpha_1,\alpha_2}_{x,y}f(x,y)$$
$$= \frac{B(\alpha_1,\alpha_2)}{(1-\alpha_1)(1-\alpha_2)} \int_a^x \int_b^y \frac{\partial^2 f(t,s)}{\partial t \partial s} \cdot E^1_{\alpha_1,\alpha_2;1,1}\left(\frac{-\alpha_1\alpha_2}{(1-\alpha_1)(1-\alpha_2)}(x-t)^{\alpha_1}(y-s)^{\alpha_2}\right) ds\, dt,$$

where $B(\alpha_1,\alpha_2)$ is the same normalisation function as above.

These definitions are valid for any $\alpha_1,\alpha_2 \in \mathbb{C}$ with positive real parts, although the main case of interest is when $\alpha_1,\alpha_2 \in (0,1)$.

Remark 3. The multiplier function $B(\alpha_1,\alpha_2)$ is introduced by analogy with the original definition of the AB derivative and AB integral [48], and the restrictions on this function are imposed in order to ensure the following limiting behaviour as the fractional orders α_1,α_2 approach 0 or 1:

$$\lim_{\alpha_1,\alpha_2 \to 0}\left(^{AB}_{a,b}I^{\alpha_1,\alpha_2}_{x,y}f(x,y)\right) = f(x,y),$$
$$\lim_{\alpha_1,\alpha_2 \to 1}\left(^{AB}_{a,b}I^{\alpha_1,\alpha_2}_{x,y}f(x,y)\right) = {}_bI^1_y {}_aI^1_x(f)(x,y),$$
$$\lim_{\alpha_1,\alpha_2 \to 0}\left(^{ABR}_{a,b}D^{\alpha_1,\alpha_2}_{x,y}f(x,y)\right) = f(x,y),$$
$$\lim_{\alpha_1,\alpha_2 \to 0}\left(^{ABC}_{a,b}D^{\alpha_1,\alpha_2}_{x,y}f(x,y)\right) = f(x,y) - f(a,y) - f(b,x) + f(a,b).$$

Note that if the pair (α_1,α_2) takes the values $(0,1)$ or $(1,0)$, then the mixed bivariate AB integral becomes zero. This is why we call it specifically a **mixed** bivariate operator: in the square set $[0,1] \times [0,1]$ for values of the pair (α_1,α_2), the values of the mixed bivariate AB integral are weighted towards the diagonal (where both the fractional orders α_1 and α_2 have more similar values) rather than towards the asymmetric corners.

The following result is the bivariate analogue of [47] (Theorem 3.1) or [11] (Proposition 2.4). It follows directly from our consideration above of the non-singular finite-sum special case of the five-parameter Mittag-Leffler kernel operators which culminated in Equations (16) and (17).

Proposition 3. The mixed bivariate AB integral and derivatives can be written in terms of the five-parameter Mittag-Leffler kernel operators as follows:

$$^{AB}_{a,b}I^{\alpha_1,\alpha_2}_{x,y}f(x,y) = \frac{(1-\alpha_1)(1-\alpha_2)}{B(\alpha_1,\alpha_2)} \cdot {}_{a,b}\mathcal{I}^{-1;\frac{-\alpha_1\alpha_2}{(1-\alpha_1)(1-\alpha_2)}}_{\alpha_1,\alpha_2;0,0}(f)(x,y)$$
$$= \frac{(1-\alpha_1)(1-\alpha_2)}{B(\alpha_1,\alpha_2)} \cdot {}_{a,b}\mathcal{D}^{1;\frac{-\alpha_1\alpha_2}{(1-\alpha_1)(1-\alpha_2)}}_{\alpha_1,\alpha_2;0,0}(f)(x,y);$$
$$^{ABR}_{a,b}D^{\alpha_1,\alpha_2}_{x,y}f(x,y) = \frac{B(\alpha_1,\alpha_2)}{(1-\alpha_1)(1-\alpha_2)} \cdot {}_{a,b}\mathcal{D}^{-1;\frac{-\alpha_1\alpha_2}{(1-\alpha_1)(1-\alpha_2)}}_{\alpha_1,\alpha_2;0,0}(f)(x,y)$$
$$= \frac{B(\alpha_1,\alpha_2)}{(1-\alpha_1)(1-\alpha_2)} \cdot {}_{a,b}\mathcal{I}^{1;\frac{-\alpha_1\alpha_2}{(1-\alpha_1)(1-\alpha_2)}}_{\alpha_1,\alpha_2;0,0}(f)(x,y);$$
$$^{ABC}_{a,b}D^{\alpha_1,\alpha_2}_{x,y}f(x,y) = \frac{B(\alpha_1,\alpha_2)}{(1-\alpha_1)(1-\alpha_2)} \cdot {}^C_{a,b}\mathcal{D}^{-1;\frac{-\alpha_1\alpha_2}{(1-\alpha_1)(1-\alpha_2)}}_{\alpha_1,\alpha_2;0,0}(f)(x,y).$$

However, due to the non-singularity properties of the AB-type operators, they can be defined on a larger function space than the general fractional derivatives with five-parameter Mittag-Leffler kernel: no differentiability assumptions are required. The following result is the bivariate analogue in [49] (Lemma 2.1) which was the first result to establish appropriate function spaces for the AB derivatives.

Theorem 8. The mixed bivariate AB integral ${}^{AB}_{a,b}I^{\alpha_1,\alpha_2}_{x,y}f(x,y)$ is defined for any function $f \in L_1([a,c] \times [b,d])$. The mixed bivariate ABR derivative ${}^{ABR}_{a,b}D^{\alpha_1,\alpha_2}_{x,y}f(x,y)$ can also be defined for any function $f \in L_1([a,c] \times [b,d])$, while its ABC counterpart ${}^{ABC}_{a,b}D^{\alpha_1,\alpha_2}_{x,y}f(x,y)$ is defined for any twice differentiable function f on $[a,c] \times [b,d]$ such that $\frac{\partial^2}{\partial x \partial y}f \in L_1([a,c] \times [b,d])$.

Proof. For the mixed bivariate AB integral, the result is clear, since this is just a linear combination of $f(x,y)$ with its double Riemann–Liouville integral.

For the mixed bivariate ABC derivative, the result follows from Theorem 2, since this operator is defined by applying a special case of the five-parameter Mittag-Leffler kernel operator to the function $\frac{\partial^2}{\partial x \partial y}f(x,y)$.

It remains to consider the mixed bivariate ABR derivative, which we can simplify in the following way since the kernel is non-singular, as a bivariate version of the arguments used in [11,49]:

$$\frac{(1-\alpha_1)(1-\alpha_2)}{B(\alpha_1,\alpha_2)} \cdot {}^{ABR}_{a,b}D^{\alpha_1,\alpha_2}_{x,y}f(x,y)$$

$$= \frac{\partial^2}{\partial x \partial y} \int_a^x \int_b^y f(t,s) E^1_{\alpha_1,\alpha_2;1,1}\left(\frac{-\alpha_1\alpha_2}{(1-\alpha_1)(1-\alpha_2)}(x-t)^{\alpha_1}(y-s)^{\alpha_2}\right) ds\, dt$$

$$= f(x,y) + \int_a^x f(t,s)\frac{\partial}{\partial x}\left[E^1_{\alpha_1,\alpha_2;1,1}\left(\frac{-\alpha_1\alpha_2}{(1-\alpha_1)(1-\alpha_2)}(x-t)^{\alpha_1}(y-s)^{\alpha_2}\right)\right]_{s=y} dt$$

$$+ \int_b^y f(t,s)\frac{\partial}{\partial y}\left[E^1_{\alpha_1,\alpha_2;1,1}\left(\frac{-\alpha_1\alpha_2}{(1-\alpha_1)(1-\alpha_2)}(x-t)^{\alpha_1}(y-s)^{\alpha_2}\right)\right]_{t=x} ds$$

$$+ \int_a^x \int_b^y f(t,s)\frac{\partial^2}{\partial x \partial y}E^1_{\alpha_1,\alpha_2;1,1}\left(\frac{-\alpha_1\alpha_2}{(1-\alpha_1)(1-\alpha_2)}(x-t)^{\alpha_1}(y-s)^{\alpha_2}\right) ds\, dt$$

$$= f(x,y) + \int_a^x \int_b^y f(t,s)\frac{\partial^2}{\partial x \partial y}E^1_{\alpha_1,\alpha_2;1,1}\left(\frac{-\alpha_1\alpha_2}{(1-\alpha_1)(1-\alpha_2)}(x-t)^{\alpha_1}(y-s)^{\alpha_2}\right) ds\, dt,$$

where the new kernel function is

$$\frac{\partial^2}{\partial x \partial y}E^1_{\alpha_1,\alpha_2;1,1}\left(\frac{-\alpha_1\alpha_2}{(1-\alpha_1)(1-\alpha_2)}(x-t)^{\alpha_1}(y-s)^{\alpha_2}\right) = \sum_{k=1}^{\infty} \frac{(x-t)^{\alpha_1 k - 1}(y-s)^{\alpha_2 k - 1}}{\Gamma(\alpha_1 k)\Gamma(\alpha_2 k)}$$

$$\sim \frac{(x-t)^{\alpha_1 - 1}(y-s)^{\alpha_2 - 1}}{\Gamma(\alpha_1)\Gamma(\alpha_2)},$$

and therefore integrable, as $x - t \to 0$ and $y - s \to 0$. Thus, the ABR derivative of $f(x,y)$ equals the function $f(x,y)$ plus an integral term which behaves as a double Riemann–Liouville integral of $f(x,y)$ in the singular limit $x - t \to 0, y - s \to 0$. This means the operator is well-defined for any $f \in L_1([a,c] \times [b,d])$, as required. □

The following result is the bivariate analogue of the original series formulae for AB derivatives, [49] (Theorems 2.1 and 2.2). It follows directly from the work we did above to obtain Equation (18).

Proposition 4. The mixed bivariate AB derivatives can be given by the following infinite series of double Riemann–Liouville fractional integrals:

$${}^{ABR}_{a,b}D^{\alpha_1,\alpha_2}_{x,y}f(x,y) = \frac{B(\alpha_1,\alpha_2)}{(1-\alpha_1)(1-\alpha_2)}\sum_{k=0}^{\infty}\left(\frac{-\alpha_1\alpha_2}{(1-\alpha_1)(1-\alpha_2)}\right)^k {}_bI^{\alpha_2 k}_y\, {}_aI^{\alpha_1 k}_x(f)(x,y),$$

$${}^{ABC}_{a,b}D^{\alpha_1,\alpha_2}_{x,y}f(x,y) = \frac{B(\alpha_1,\alpha_2)}{(1-\alpha_1)(1-\alpha_2)}\sum_{k=0}^{\infty}\left(\frac{-\alpha_1\alpha_2}{(1-\alpha_1)(1-\alpha_2)}\right)^k {}_bI^{\alpha_2 k + 1}_y\, {}_aI^{\alpha_1 k + 1}_x\left(\frac{\partial^2 f}{\partial x \partial y}\right)(x,y).$$

where this series is locally uniformly convergent for any $f \in L_1([a,c] \times [b,d])$.

Corollary 1. *From the series formulae above, it is clear that the mixed bivariate ABC derivative can be written in terms of its ABR counterpart as follows:*

$$^{ABC}_{a,b}D^{\alpha_1,\alpha_2}_{x,y}f(x,y) = {}^{ABR}_{a,b}D^{\alpha_1,\alpha_2}_{x,y}\Big[f(x,y) - f(a,y) - f(b,x) + f(a,b)\Big].$$

Therefore, the domain of this operator can be extended from the one mentioned in Theorem 8, and the mixed bivariate ABC derivative can also be defined on the whole space $L_1([a,c] \times [b,d])$, without any differentiability conditions.

Theorem 9. *The mixed bivariate AB integral is both a left and right inverse of the mixed bivariate ABR derivative on the space $L_1([a,c] \times [b,d])$, and it shares the following relationship with the mixed bivariate ABC derivative:*

$$^{AB}_{a,b}I^{\alpha_1,\alpha_2}_{x,y} \, {}^{ABC}_{a,b}D^{\alpha_1,\alpha_2}_{x,y}f(x,y) = f(x,y) - f(a,y) - f(b,x) + f(a,b).$$

Proof. This is immediate from Proposition 4 and Corollary 1. Note that the series for the ABR derivative consists only of Riemann–Liouville fractional *integrals*, which have a semigroup property unlike their fractional derivative counterparts. □

The following result is the bivariate analogue of the Laplace transform for AB derivatives [48,49].

Proposition 5. *The double Laplace transforms of the mixed bivariate AB integral and derivatives, in the case $a = b = 0$, can be expressed as follows:*

$$\mathcal{L}_2\Big[{}^{AB}_{0,0}I^{\alpha_1,\alpha_2}_{x,y}f(x,y)\Big](p,q) = \frac{(1-\alpha_1)(1-\alpha_2)}{B(\alpha_1,\alpha_2)}\left(1 + \frac{\alpha_1\alpha_2}{(1-\alpha_1)(1-\alpha_2)}p^{-\alpha_1}q^{-\alpha_2}\right)\mathcal{L}_2[f](p,q);$$

$$\mathcal{L}_2\Big[{}^{ABR}_{0,0}D^{\alpha_1,\alpha_2}_{x,y}f(x,y)\Big](p,q) = \frac{B(\alpha_1,\alpha_2)}{(1-\alpha_1)(1-\alpha_2)}\left(1 + \frac{\alpha_1\alpha_2}{(1-\alpha_1)(1-\alpha_2)}p^{-\alpha_1}q^{-\alpha_2}\right)^{-1}\mathcal{L}_2[f](p,q);$$

$$\mathcal{L}_2\Big[{}^{ABC}_{0,0}D^{\alpha_1,\alpha_2}_{x,y}f(x,y)\Big](p,q) = \frac{B(\alpha_1,\alpha_2)}{(1-\alpha_1)(1-\alpha_2)}\left(1 + \frac{\alpha_1\alpha_2}{(1-\alpha_1)(1-\alpha_2)}p^{-\alpha_1}q^{-\alpha_2}\right)^{-1}$$
$$\times \left(\mathcal{L}_2[f](p,q) - \frac{1}{p}\mathcal{L}_{y\to q}[f](0,q) - \frac{1}{q}\mathcal{L}_{x\to p}[f](p,0) + \frac{1}{pq}f(0,0)\right).$$

Proof. For the mixed bivariate AB integral, the result follows directly from the definition and the known facts on Laplace transforms of Riemann–Liouville fractional integrals.

For the mixed bivariate AB derivative of Riemann–Liouville type, the result follows from the series formula of Proposition 4 and a simple application of the binomial theorem. An important fact to notice here is that, via the series formula, this Riemann–Liouville type operator can be written solely in terms of Riemann–Liouville *integrals*, so there is no need to involve initial conditions here.

For the mixed bivariate AB derivative of Caputo type, derivatives and therefore initial conditions become involved. We use the correct form for the double Laplace transform of a mixed partial derivative, given in [46,50], to achieve the result:

$$\mathcal{L}_2\Big[{}^{ABC}_{a,b}D^{\alpha_1,\alpha_2}_{x,y}f(x,y)\Big](p,q)$$
$$= \frac{B(\alpha_1,\alpha_2)}{(1-\alpha_1)(1-\alpha_2)}\sum_{k=0}^{\infty}\left(\frac{-\alpha_1\alpha_2}{(1-\alpha_1)(1-\alpha_2)}\right)^k q^{-\alpha_2 k - 1}p^{-\alpha_1 k - 1}\mathcal{L}_2\left[\frac{\partial^2 f}{\partial x \partial y}(x,y)\right](p,q)$$
$$= \frac{B(\alpha_1,\alpha_2)}{(1-\alpha_1)(1-\alpha_2)}\left(1 + \frac{\alpha_1\alpha_2}{(1-\alpha_1)(1-\alpha_2)}p^{-\alpha_1}q^{-\alpha_2}\right)^{-1}p^{-1}q^{-1}\Big(pq\mathcal{L}_2[f](p,q)$$
$$- q\mathcal{L}_{y\to q}[f](0,q) - p\mathcal{L}_{x\to p}[f](p,0) + f(0,0)\Big),$$

which trivially rearranges to the stated result. □

5. Discussion and Conclusions

In this paper, we construct a type of bivariate fractional calculus based originally on a function of a single variable defined by a single power series. We prove some of the fundamental properties of this type of fractional calculus: boundedness of operators, series formulae, Laplace transforms, semigroup and inversion properties, etc. As a special case, we also consider the non-singular versions of our bivariate operators, in order to construct a two-dimensional version of the well-known Atangana–Baleanu calculus.

Some of the ideas used in the above work are similar to those for previously existing models of fractional calculus with various Mittag-Leffler type kernels; however, this is the first time that a univariate function has been used in this way to construct bivariate fractional-calculus operators. Previous contributions in this direction have included using univariate single series to construct univariate operators [10], using bivariate double series to construct univariate operators [18], and using bivariate double series to construct bivariate operators [23,24].

At first glance, it may seem that our construction is unnatural: Given a function of one variable defined by a single series, why would one make a simple thing more complicated by replacing z with $x^{\alpha_1}y^{\alpha_2}$ and introducing a bivariate integral operator? The answer is that the underlying mathematical structure is still simple, because everything is defined by a single series whose convergence is easy to describe, but the range of problems that can be tackled is now richer and more diverse, because more variables and parameters allow for more flexibility in adapting the operators to particular scenarios. Our "trick" of turning a univariate function into a bivariate operator gives a shortcut, a possibility of modelling complicated problems using simpler tools.

The bivariate operators that we define lend themselves more to partial differential equations than to ordinary differential equations. Even our initial motivation for proposing them, in the first section of this paper, arose from a bivariate Abel-type equation for a function of two variables. In the literature thus far, Mittag-Leffler kernel operators have mostly been applied to modelling problems with ordinary differential equations. While the existing operators could of course be combined to trivially create bivariate derivatives and integrals with Mittag-Leffler behaviour, we believe that a richer structure will emerge from considering operators such as those here, where the behaviours with respect to x and y are intertwined so that the operator cannot simply be broken down into one with respect to x and another with respect to y. We believe that our work here may have useful ramifications in the study of fractional partial differential equations in 2 or 2 + 1 dimensions.

For higher dimensions, it will be possible to construct similarly a model of trivariate or multivariate fractional calculus based on a univariate function defined by a single series with 3 or n different gamma functions on the denominator. We give here, without justification, the trivariate version.

Starting from the following function, defined by a series convergent for parameters $\alpha_i, \beta_i, \gamma \in \mathbb{C}$ ($i = 1, 2, 3$) with $\text{Re}(\alpha_1 + \alpha_2 + \alpha_3) > 0$:

$$E^{\gamma}_{\alpha_1,\alpha_2,\alpha_3;\beta_1,\beta_2,\beta_3}(z) = \sum_{k=0}^{\infty} \frac{(\gamma)_k}{\Gamma(\alpha_1 k + \beta_1)\Gamma(\alpha_2 k + \beta_2)\Gamma(\alpha_3 k + \beta_k)} \cdot \frac{z^k}{k!},$$

it is possible to define the following trivariate fractional integral operator, convergent for any function $f \in L_1([a_1, b_1] \times [a_2, b_2] \times [a_3, b_3])$ and parameters $\alpha_i, \beta_i, \gamma, \lambda \in \mathbb{C}$ with $\text{Re}(\alpha_i) > 0$ and $\text{Re}(\beta_i) > 0$ for $i = 1, 2, 3$:

$$_{a_1,a_2,a_3}\mathcal{I}^{\gamma;\lambda}_{\alpha_1,\alpha_2,\alpha_3;\beta_1,\beta_2,\beta_3}(f)(x,y)$$
$$:= \int_{a_1}^{x} \int_{a_2}^{y} \int_{a_3}^{z} f(s,t,u)(x-s)^{\beta_1-1}(y-t)^{\beta_2-1}(z-u)^{\beta_3-1}$$
$$\times E^{\gamma}_{\alpha_1,\alpha_2,\alpha_3;\beta_1,\beta_2,\beta_3}\big(\lambda(x-s)^{\alpha_1}(y-t)^{\alpha_2}(z-u)^{\alpha_3}\big)\,du\,dt\,ds,$$

and then the corresponding trivariate fractional derivative operator are defined as follows for parameters $\alpha_i, \beta_i, \gamma, \lambda \in \mathbb{C}$ with $\text{Re}(\alpha_i) > 0$ and $\text{Re}(\beta_i) \geq 0$ for $i = 1, 2, 3$:

$$_{a_1,a_2,a_3}\mathcal{D}^{\gamma;\lambda}_{\alpha_1,\alpha_2,\alpha_3;\beta_1,\beta_2,\beta_3}(f)(x,y) = \frac{\partial^{N_1+N_2+N_3}}{\partial x^{N_1} \partial y^{N_2} \partial z^{N_3}} \left(_{a_1,a_2,a_3}\mathcal{D}^{-\gamma;\lambda}_{\alpha_1,\alpha_2,\alpha_3;N_1-\beta_1,N_2-\beta_2,N_3-\beta_3} f\right)(x,y).$$

The detailed analysis and investigation of these functions and operators, including verification of their essential properties and discussion of special cases and applications, is left for a future research project.

Other related directions of research may include a deeper study of the function spaces on which the integral and derivative operators considered here can be defined. For example, Theorem 2 can be extended easily to show that the bivariate fractional integral operator is bounded on an L^p space. Then, a measure-theoretical approach may yield extensions to Morrey spaces [51], or a distributional approach may yield extensions to still larger spaces related to generalised integral operators [52]. Describing more deeply the various relevant function spaces may be useful in the qualitative theory of partial differential equations related to these operators, for example regularity theory [29].

Author Contributions: All authors contributed equally to this article. All authors have read and agreed to the published version of the manuscript.

Funding: This research received no external funding.

Data Availability Statement: Not applicable.

Conflicts of Interest: The authors declare no conflict of interest.

References

1. Gorenflo, R.; Kilbas, A.A.; Mainardi, F.; Rogosin, S.V. *Mittag-Leffler Functions, Related Topics and Applications*, 2nd ed.; Springer: Berlin, Germany, 2020.
2. Paneva-Konovska, J. *From Bessel to Multi-Index Mittag-Leffler Functions: Enumerable Families, Series in Them and Convergence*; World Scientific: Singapore, 2016.
3. Kilbas, A.A.; Srivastava, H.M.; Trujillo, J.J. *Theory and Applications of Fractional Differential Equations*; Elsevier: Amsterdam, The Netherlands, 2006.
4. Shishkina, E.; Sitnik, S. *Transmutations, Singular and Fractional Differential Equations with Applications to Mathematical Physics*; Academic Press: Cambridge, MA, USA, 2020.
5. Mathai, A.M.; Haubold, H.J. Mittag-Leffler Functions and Fractional Calculus. In *Special Functions for Applied Scientists*; Mathai, A.M., Haubold, H.J., Eds.; Springer: Berlin, Germany, 2008; pp. 79–134.
6. Samko, S.G.; Kilbas, A.A.; Marichev, O.I. *Fractional Integrals and Derivatives: Theory and Applications*; Gordon & Breach Science Publishers: Yverdon, Switzerland, 1993; Originally in Russian: Nauka i Tekhnika, Minsk, 1987.
7. Baleanu, D.; Diethelm, K.; Scalas, E.; Trujillo, J.J. *Fractional Calculus: Models and Numerical Methods*, 2nd ed.; World Scientific: New York, NY, USA, 2017.
8. Hilfer, R. (Ed.) *Applications of Fractional Calculus in Physics*; World Scientific: Singapore, 2000.
9. Rabotnov, Y.N. *Elements of Hereditary Solid Mechanics*; Mir Publishers: Moscow, Russia, 1980.
10. Prabhakar, T.R. A singular integral equation with a generalized Mittag Leffler function in the kernel. *Yokohama Math. J.* **1971**, *19*, 7–15.
11. Fernandez, A.; Baleanu, D. Classes of Operators in Fractional Calculus: A Case Study. *Math. Methods Appl. Sci.* **2020**, 1–20. [CrossRef]
12. Fernandez, A.; Baleanu, D.; Srivastava, H.M. Series representations for models of fractional calculus involving generalised Mittag-Leffler functions. *Commun. Nonlinear Sci. Numer. Simul.* **2019**, *67*, 517–527. [CrossRef]
13. Kilbas, A.A.; Saigo, M.; Saxena, R.K. Generalized Mittag-Leffler function and generalized fractional calculus operators. *Integral Transform. Spec. Funct.* **2004**, *15*, 31–49. [CrossRef]
14. Hadid, S.B.; Luchko, Y.F. An operational method for solving fractional differential equations of an arbitrary real order. *Panam. Math. J.* **1996**, *6*, 57–73.
15. Luchko, Y.F.; Gorenflo, R. An operational method for solving fractional differential equations. *Acta Math. Vietnam.* **1999**, *24*, 207–234.
16. Saxena, R.K.; Kalla, S.L.; Saxena, R. Multivariate analogue of generalised Mittag-Leffler function. *Integral Transform. Spec. Funct.* **2011**, *22*, 533–548. [CrossRef]
17. Özarslan, M.A.; Fernandez, A. On the fractional calculus of multivariate Mittag-Leffler functions. *Int. J. Comput. Math.* **2021**, 1–27. [CrossRef]

18. Fernandez, A.; Kürt, C.; Özarslan, M.A. A naturally emerging bivariate Mittag-Leffler function and associated fractional-calculus operators. *Comput. Appl. Math.* **2020**, *39*, 200. [CrossRef]
19. Huseynov, I.T.; Ahmadova, A.; Ojo, G.O.; Mahmudov, N.I. A natural extension of Mittag-Leffler function associated with a triple infinite series. *arXiv* **2020**, arXiv:2011.03999.
20. Huseynov, I.T.; Ahmadova, A.; Fernandez, A.; Mahmudov, N.I. Explicit analytic solutions of incommensurate fractional differential equation systems. *Appl. Math. Comput.* **2021**, *390*, 125590.
21. Ahmadova, A.; Huseynov, I.T.; Fernandez, A.; Mahmudov, N.I. Trivariate Mittag-Leffler functions used to solve multi-order systems of fractional differential equations. *Commun. Nonlinear Sci. Numer. Simul.* **2021**, *97*, 105735. [CrossRef]
22. Garg, M.; Manohar, P.; Kalla, S.L. A Mittag-Leffler-type function of two variables. *Integral Transform. Spec. Funct.* **2013**, *24*, 934–944. [CrossRef]
23. Özarslan, M.A.; Kürt, C. On a double integral equation including a set of two variables polynomials suggested by Laguerre polynomials. *J. Comput. Anal. Appl.* **2017**, *22*, 1198–1207.
24. Özarslan, M.A.; Kürt, C. Bivariate Mittag-Leffler functions arising in the solutions of convolution integral equation with 2D-Laguerre-Konhauser polynomials in the kernel. *Appl. Math. Comput.* **2019**, *347*, 631–644. [CrossRef]
25. Kiryakova, V. The multi-index Mittag-Leffler functions as an important class of special functions of fractional calculus. *Comput. Math. Appl.* **2010**, *59*, 1885–1895. [CrossRef]
26. Kiryakova, V. Unified approach to fractional calculus images of special functions—A survey. *Mathematics* **2020**, *8*, 2260. [CrossRef]
27. Kiryakova, V. A guide to special functions in fractional calculus. *Mathematics* **2021**, *9*, 106. [CrossRef]
28. Fernandez, A.; Baleanu, D.; Fokas, A.S. Solving PDEs of fractional order using the unified transform method. *Appl. Math. Comput.* **2018**, *339*, 738–749. [CrossRef]
29. Fernandez, A. An elliptic regularity theorem for fractional partial differential operators. *Comput. Appl. Math.* **2018**, *37*, 5542–5553. [CrossRef]
30. Williams, P. Fractional Calculus of Schwartz Distributions. Bachelor's Thesis, University of Melbourne, Melbourne, Australia, 2007.
31. Djida, J.-D.; Area, I.; Nieto, J.J. Nonlocal time porous medium equation with fractional time derivative. *Rev. Mat. Complut.* **2019**, *32*, 273–304. [CrossRef]
32. Djida, J.-D.; Area, I.; Nieto, J.J. Nonlocal time-porous medium equation: Weak solutions and finite speed of propagation. *Discret. Contin. Dyn. Syst. Ser. B* **2019**, *24*, 4031–4053. [CrossRef]
33. Lischke, A.; Pang, G.; Gulian, M.; Song, F.; Glusa, C.; Zheng, X.; Mao, Z.; Cai, W.; Meerschaert, M.M.; Ainsworth, M.; et al. What is the fractional Laplacian? A comparative review with new results. *J. Comput. Phys.* **2020**, *404*, 109009. [CrossRef]
34. Stinga, P.R. User's guide to the fractional Laplacian and the method of semigroups. *arXiv* **2018**, arXiv:1808.05159.
35. Conlan, J. Hyperbolic Differential Equations of Generalized Order. *Appl. Anal.* **1983**, *14*, 167–177. [CrossRef]
36. Jiang, J.; Feng, Y.; Li, S. Exact Solutions to the Fractional Differential Equations with Mixed Partial Derivatives. *Axioms* **2018**, *7*, 10. [CrossRef]
37. Kilbas, A.A.; Repin, O.A. An Analog of the Tricomi Problem for a Mixed Type Equation with a Partial Fractional Derivative. *Fract. Calc. Appl. Anal.* **2010**, *13*, 69–84.
38. Kilbas, A.A.; Saigo, M.; Trujillo, J.J. On the generalized Wright function. *Fract. Calc. Appl. Anal.* **2002**, *5*, 437–460.
39. Fernandez, A.; Husain, I. Modified Mittag-Leffler functions with applications in complex formulae for fractional calculus. *Fractal Fract.* **2020**, *4*, 45. [CrossRef]
40. Luchko, Y. The Four-Parameters Wright Function of the Second kind and its Applications in FC. *Mathematics* **2020**, *8*, 970. [CrossRef]
41. Al-Bassam, M.A.; Luchko, Y.F. On generalized fractional calculus and its application to the solution of integro-differential equations. *J. Fract. Calc.* **1995**, *7*, 69–88.
42. Choi, J.; Agarwal, P. A note on fractional integral operator associated with multiindex Mittag-Leffler functions. *Filomat* **2016**, *30*, 1931–1939. [CrossRef]
43. Marichev, O.I. *Handbook of Integral Transforms of Higher Transcendental Functions, Theory and Algorithmic Tables*; Ellis Horwood: Chichester, UK, 1983.
44. Miller, K.S.; Ross, B. *An Introduction to the Fractional Calculus and Fractional Differential Equations*; Wiley: New York, NY, USA, 1993.
45. Oldham, K.B.; Spanier, J. *The Fractional Calculus*; Academic Press: San Diego, CA, USA, 1974.
46. Kürt, C.; Özarslan, M.A.; Fernandez, A. On a certain bivariate Mittag-Leffler function analysed from a fractional-calculus point of view. *Math. Methods Appl. Sci.* **2021**, *44*, 2600–2620. [CrossRef]
47. Fernandez, A.; Abdeljawad, T.; Baleanu, D. Relations between fractional models with three-parameter Mittag-Leffler kernels. *Adv. Differ. Equations* **2020**, *2020*, 186. [CrossRef]
48. Atangana, A.; Baleanu, D. New fractional derivative with non-local and non-singular kernel. *Therm. Sci.* **2016**, *20*, 757–763. [CrossRef]
49. Baleanu, D.; Fernandez, A. On some new properties of fractional derivatives with Mittag-Leffler kernel. *Commun. Nonlinear Sci. Numer. Simul.* **2018**, *59*, 444–462. [CrossRef]
50. Anwar, A.M.O.; Jarad, F.; Baleanu, D.; Ayaz, F. Fractional Caputo heat equation within the double Laplace transform. *Rom. J. Phys.* **2013**, *58*, 15–22.

51. Eridani, A.; Kokilashvili, V.; Meskhi, A. Morrey spaces and fractional integral operators. *Expo. Math.* **2009**, *27*, 227–239. [CrossRef]
52. Morales, M.G.; Došla, Z.; Mendoza, F.J. Riemann–Liouville derivative over the space of integrable distributions. *Electron. Res. Arch.* **2020**, *28*, 567–587. [CrossRef]

fractal and fractional

Article

Solving a System of Fractional-Order Volterra-Fredholm Integro-Differential Equations with Weakly Singular Kernels via the Second Chebyshev Wavelets Method

Esmail Bargamadi [1], Leila Torkzadeh [1,*], Kazem Nouri [1] and Amin Jajarmi [2,*]

1. Department of Mathematics, Faculty of Mathematics, Statistics and Computer Sciences, Semnan University, P.O. Box 35195-363, Semnan 35131-19111, Iran; esmailbargamadi@semnan.ac.ir (E.B.); knouri@semnan.ac.ir (K.N.)
2. Department of Electrical Engineering, University of Bojnord, P.O. Box 94531-1339, Bojnord 94531-55111, Iran
* Correspondence: torkzadeh@semnan.ac.ir (L.T.); a.jajarmi@ub.ac.ir (A.J.); Tel.: +98-2331-535-768 (L.T.); +98-5832-201-000 (A.J.)

Abstract: In this paper, by means of the second Chebyshev wavelet and its operational matrix, we solve a system of fractional-order Volterra–Fredholm integro-differential equations with weakly singular kernels. We estimate the functions by using the wavelet basis and then obtain the approximate solutions from the algebraic system corresponding to the main system. Moreover, the implementation of our scheme is presented, and the error bounds of approximations are analyzed. Finally, we evaluate the efficiency of the method through a numerical example.

Keywords: second Chebyshev wavelet; system of Volterra–Fredholm integro-differential equations; fractional-order Caputo derivative operator; fractional-order Riemann–Liouville integral operator; error bound

MSC: 34A08; 26A33; 65T60; 45F15

1. Introduction

Fractional calculus has been the interest of many scientists and engineers [1–3]. Many engineering and science phenomena, such as the heat conduction problem, radiative equilibrium, elasticity and fracture mechanics [4], viscoelastic deformation, viscoelasticity, viscous fluid [5], continuous population [6] and so forth, are modeled using the fractional integro-differential equations with a weakly singular kernel, fractional differential equations, fractional integral equations and system of nonlinear Volterra integro-differential equations. Many applied problems are transformed into the system of fractional differential and integral equations by mathematical modeling [7–10]. Consequently, it is essential to obtain the approximate solution of a system of integro-differential equations by numerical methods.

In this paper, we solve a system of fractional-order Volterra–Fredholm integro-differential equations with weakly singular kernels in the following form:

$$\begin{cases} D^{\alpha_1} y_1(t) = \lambda_1 \int_0^t \frac{y_1(s)}{(t-s)^{\beta_1}} ds + \lambda_2 \int_0^1 k_1(t,s) y_2(s) ds + f_1(t), \\ D^{\alpha_2} y_2(t) = \lambda_3 \int_0^t \frac{y_1(s)}{(t-s)^{\beta_2}} ds + \lambda_4 \int_0^1 k_2(t,s) y_2(s) ds + f_2(t), \end{cases} \quad y_1(0) = a_1, y_2(0) = a_2, \quad (1)$$

where $y_1(t), y_2(t)$ are unknown functions, the functions $f_1(t), f_2(t), k_1(t,s)$, and $k_2(t,s)$ are known, and $\lambda_1, \lambda_2, \lambda_3, \lambda_4, a_1, a_2$ are real constants, where $0 < \alpha_1, \alpha_2, \beta_1, \beta_2 < 1$ and $D^{\alpha_1}, D^{\alpha_2}$ denote the Caputo fractional-order derivatives. Furthermore, $\frac{1}{(t-s)^{\beta_1}}$ and $\frac{1}{(t-s)^{\beta_2}}$ are the weakly singular kernels of the system of fractional-order Volterra–Fredholm integro-differential equations.

In recent years, researchers have proposed different methods for solving the system of differential equations. In 2015, Sahu and Ray [11] developed a numerical method based on the Legendre hybrid block pulse function to approximate the solution of nonlinear systems of Fredholm-Manhattan integral equations. In the same year, they presented another scheme to solve a system of nonlinear Volterra integro-differential equations using Legendre wavelets [6]. Yüzbasi [12] solved the system of linear Fredholm–Volterra integro-differential equations, which includes the derivatives of unknown functions in integral parts using the Bessel collocation method. In 2016, Deif and Grace [13] developed a new iterative method to approximate the solution of a system of linear fractional differential integral equations. In 2019, Xie and Yi [14] developed a numerical method for solving a nonlinear system of fractional-order Volterra–Fredholm integro-differential equations based on block-pulse functions. In 2020, Saemi et al. [15] developed a solution for the system of fractional-order Volterra–Fredholm integro-differential equations based on Müntz–Legendre wavelets.

Wavelets are one of the most important tools used in various fields such as quantum mechanics, signal processing, image processing, time-frequency analysis and data compression [16]. One of the methods that have been considered in recent years to solve various ordinary and fractional-order equations is the use of wavelets [17–19]. Wavelets provide a detailed accurate representation of different types of functions and operators, and their relationship with numerical algorithms [20,21]. The second Chebyshev wavelet is one of the wavelets, which has gained attention in solving many problems and is applicable for solving various types of Volterra integral equations with a weakly singular kernel [17], fractional-order nonlinear Fredholm integro-differential equations [20], fractional-order differential equations [22], a system of linear differential equations [23], fractional-order integro-differential equations with a weakly singular kernel [5], and Abel's integral equations [16]. Approximation of equations using the second Chebyshev wavelets has been considered by many researchers, such as Zhu and Wang [17,24], Zhou and Xu [21], Wang and Fan [22], Tavassoli Kajani et al. [25], Zhou et al. [26], Yi et al. [27], Lal and Sharma [16], and Manchanda and Rani [23].

In this paper, we apply the second Chebyshev wavelets method to solve the system of fractional-order Volterra–Fredholm integro-differential equations with a weakly singular kernel. In fact, the main purpose of this study is solving the system of equations with singularity. The second Chebyshev wavelets method converts the system of fractional-order Volterra–Fredholm integro-differential equations with a weakly singular kernel to a system of algebraic equations, which can be solved using the conventional linear methods.

2. Preliminaries

In this section, we introduce fractional-order operators, block-pulse functions and explain their features.

Definition 1. *The Riemann–Liouville fractional-order integral operator of order α is given by [17,27]:*

$$I^\alpha f(t) = \frac{1}{\Gamma(\alpha)} \int_a^t (t-\tau)^{\alpha-1} f(\tau) d\tau, \quad \alpha > 0,$$

where $\alpha \in (m-1, m], m \in \mathbb{N}$.

The properties of this operator are as follows:

1. $I_a^\alpha I_a^\beta f = I_a^{\alpha+\beta} f,$
2. $I_a^\alpha I_a^\beta f = I_a^\beta I_a^\alpha f,$
3. $I_a^\alpha t^c = \frac{\Gamma(c+1)}{\Gamma(c+\alpha+1)} t^{c+\alpha}.$

Definition 2. *The Caputo fractional derivative operator of order α for a function f is given by [17,18,27]:*

$$D^\alpha f(t) = \frac{1}{\Gamma(n-\alpha)} \int_a^t \frac{f^{(n)}(\tau)}{(t-\tau)^{\alpha-n+1}} d\tau, \quad \alpha \in (n-1, n],$$

where $n \in \mathbb{N}$, and $\Gamma(.)$ is the Gamma function.

The properties between the Caputo fractional-order derivative operator and the Riemann–Liouville fractional-order integral operator is given by the following expressions [17,18,24]:

1.
$$D^\alpha I^\alpha f(t) = f(t),$$

2.
$$I^\alpha D^\alpha f(t) = f(t) - \sum_{k=0}^{m-1} \frac{f^{(k)}(0)}{k!} t^k. \quad (2)$$

Definition 3. *The set of block pulse function on $[0, 1)$ is defined as:*

$$b_i(t) = \begin{cases} 1, & \frac{i-1}{m} \leq t < \frac{i}{m} \\ 0, & otherwise \end{cases}$$

where $i = 1, 2, \ldots, m$. Furthermore, the vector of block pulse functions is obtained as follows:

$$B_m(t) = [b_1(t), b_2(t), ..., b_m(t)]^T,$$

and the important properties of these functions are as follows:

1. $b_i(t)b_j(t) = \begin{cases} b_i(t), & i = j, \\ 0, & i \neq j, \end{cases}$

2. $\int_0^1 b_i(t)b_j(t)dt = \begin{cases} \frac{1}{m}, & i = j, \\ 0, & i \neq j. \end{cases}$

Lemma 1. *The block pulse function operational matrix of fractional-order integration F^α is obtained by:*

$$I^\alpha(B_m(t)) \approx F^\alpha B_m(t),$$

where

$$F^\alpha = \frac{1}{m^\alpha} \frac{1}{\Gamma(\alpha+2)} \begin{bmatrix} 1 & \xi_1 & \xi_2 & \xi_3 & \cdots & \xi_{m-1} \\ 0 & 1 & \xi_1 & \xi_2 & \cdots & \xi_{m-2} \\ 0 & 0 & 1 & \xi_1 & \cdots & \xi_{m-3} \\ \vdots & \vdots & \ddots & \ddots & \ddots & \vdots \\ 0 & 0 & \cdots & 0 & 1 & \xi_1 \\ 0 & 0 & 0 & \cdots & 0 & 1 \end{bmatrix},$$

and $\xi_k = (k+1)^{\alpha+1} - 2k^{\alpha+1} + (k-1)^{\alpha+1}$, $k = 1, 2, \ldots, m-1$.

For example with $\alpha = 0.5$ and $m = 6$:

$$F^{0.5} = \begin{bmatrix} 0.3071 & 0.2544 & 0.1656 & 0.1339 & 0.1156 & 0.1033 \\ 0 & 0.3071 & 0.2544 & 0.1656 & 0.1339 & 0.1156 \\ 0 & 0 & 0.3071 & 0.2544 & 0.1656 & 0.1339 \\ 0 & 0 & 0 & 0.3071 & 0.2544 & 0.1656 \\ 0 & 0 & 0 & 0 & 0.3071 & 0.2544 \\ 0 & 0 & 0 & 0 & 0 & 0.3071 \end{bmatrix}.$$

3. The Second Chebyshev Wavelets and Function Approximation

In this section, we introduce the second Chebyshev wavelet and then use this basis to provide an approximation of functions.

3.1. The Second Chebyshev Wavelets and Their Properties

In this part, we introduce the second Chebyshev wavelet and its features.

Definition 4. *The second Chebyshev wavelet on the interval $[0,1)$ is defined as [16,23,27]:*

$$\psi_{nm}(t) = \begin{cases} 2^{\frac{k}{2}}\sqrt{\frac{2}{\pi}}U_m(2^k t - 2n + 1), & \frac{n-1}{2^{k-1}} \leq t < \frac{n}{2^{k-1}}, \\ 0, & \text{otherwise,} \end{cases}$$

where $n = 1, 2, \ldots, 2^{k-1}$, $m = 0, 1, \ldots, M-1$, and $k, M \in \mathbb{N}$. The coefficient $\sqrt{\frac{2}{\pi}}$ is for orthonormality, and $U_m(t)$ is the Chebyshev polynomial of the second kind with degree m, which is as follows:

$$U_0(t) = 1, \quad U_1(t) = 2t, \quad U_{m+1}(t) = 2tU_m(t) - U_{m-1}(t).$$

Furthermore, the weight function of the second kind Chebyshev polynomials is $\omega(t) = \sqrt{1-t^2}$, and with transmission and dilation, first we obtain $\hat{\omega}(t) = \omega(2t-1)$, and then we get $\omega_n(t) = \omega(2^k t - 2n + 1)$ as the weight function of the second Chebyshev wavelets basis.

For example, with $k = 2$ and $M = 3$, we have $n = 1, 2$, $m = 0, 1, 2$, and for $0 \leq t < 0.5$,

$$\psi_{10}(t) = 2\sqrt{\frac{2}{\pi}}U_0(4t - 1) = 2\sqrt{\frac{2}{\pi}},$$

$$\psi_{11}(t) = 2\sqrt{\frac{2}{\pi}}U_1(4t - 1) = 2\sqrt{\frac{2}{\pi}}(8t - 2),$$

$$\psi_{12}(t) = 2\sqrt{\frac{2}{\pi}}U_2(4t - 1) = 2\sqrt{\frac{2}{\pi}}(64t^2 - 32t + 3),$$

and also for $0.5 \leq t < 1$,

$$\psi_{20}(t) = 2\sqrt{\frac{2}{\pi}}U_0(4t - 3) = 2\sqrt{\frac{2}{\pi}},$$

$$\psi_{21}(t) = 2\sqrt{\frac{2}{\pi}}U_1(4t - 3) = 2\sqrt{\frac{2}{\pi}}(8t - 6,)$$

$$\psi_{22}(t) = 2\sqrt{\frac{2}{\pi}}U_2(4t - 3) = 2\sqrt{\frac{2}{\pi}}(64t^2 - 96t + 35).$$

The second Chebyshev wavelets have an orthonormal basis of $L^2[0,1)$, i.e.,

$$<\psi_{nm}(t), \psi_{n'm'}(t)>_{\omega_n} = \int_0^1 \psi_{nm}(t)\psi_{n'm'}(t)\omega_n(t)dt = \begin{cases} 1, & m = m', n = n', \\ 0, & \text{o.w.,} \end{cases}$$

where $<.,.>_{\omega_n}$ denotes the inner product. The second Chebyshev wavelet has compact support $[\frac{n-1}{2^{k-1}}, \frac{n}{2^{k-1}}], n = 1, \ldots, 2^{k-1}$. The Chebyshev wavelet charts for $k = 3$ and $M = 4$ are shown in Figure 1.

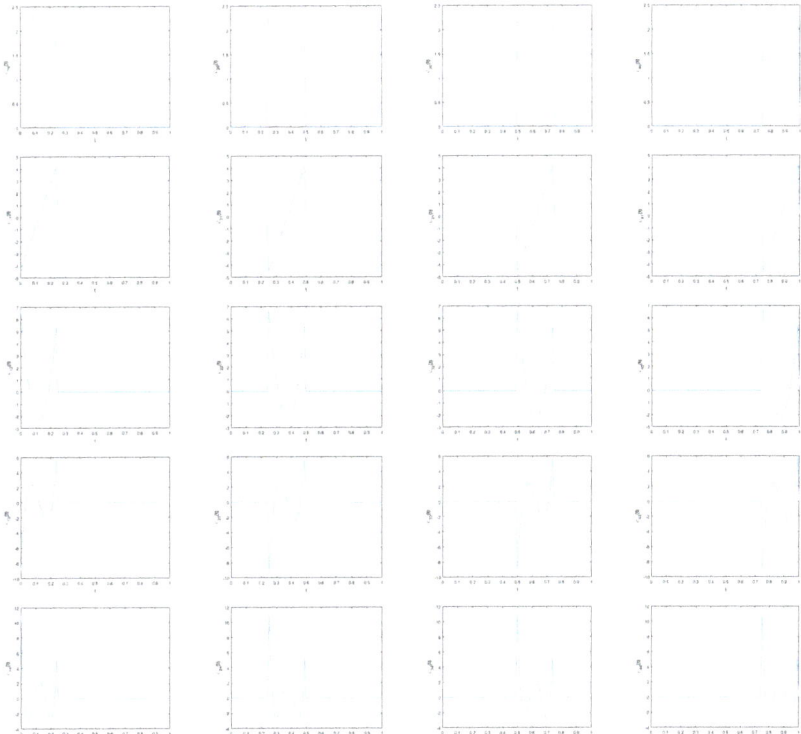

Figure 1. The second Chebyshev wavelet charts for $k = 3$ and $M = 4$.

According to the second Chebyshev wavelet, the vector of this wavelet is given by [16,27]:

$$\Psi(t) = [\psi_{10}(t), \psi_{11}(t), \ldots, \psi_{1(M-1)}(t), \\ \psi_{20}(t), \psi_{21}(t), \ldots, \psi_{2(M-1)}(t), \ldots, \\ \psi_{2^{k-1}0}(t), \psi_{2^{k-1}1}(t), \ldots, \psi_{2^{k-1}(M-1)}(t)]^T. \tag{3}$$

In other words, for $0 \leq t < 0.5$:

$$\Psi(t) = \begin{bmatrix} 2\sqrt{\frac{2}{\pi}} \\ 2\sqrt{\frac{2}{\pi}}(8t-2) \\ 2\sqrt{\frac{2}{\pi}}(64t^2 - 32t + 3) \\ 0 \\ 0 \\ 0 \end{bmatrix},$$

and for $0.5 \leq t < 1$:

$$\Psi(t) = \begin{bmatrix} 0 \\ 0 \\ 0 \\ 2\sqrt{\frac{2}{\pi}} \\ 2\sqrt{\frac{2}{\pi}}(8t-6) \\ 2\sqrt{\frac{2}{\pi}}(64t^2 - 96t + 35) \end{bmatrix}.$$

Moreover, for the collocation points [24]

$$t_i = \frac{2i-1}{2m'}, \quad i = 1, 2, \ldots, m',$$

with $m' = 2^{k-1}M$, the second Chebyshev wavelets matrix is obtained as follows [24,25]:

$$\Phi_{m' \times m'} = [\Psi(\frac{1}{2m'}), \Psi(\frac{3}{2m'}), \ldots, \Psi(\frac{2m'-1}{2m'})].$$

For $k = 2$ and $M = 3$, i.e.,

$$\Phi = \begin{bmatrix} 1.59576 & 1.59576 & 1.59576 & 0 & 0 & 0 \\ -2.12769 & 0 & 2.12769 & 0 & 0 & 0 \\ 1.24115 & -1.59576 & 1.24115 & 0 & 0 & 0 \\ 0 & 0 & 0 & 1.59576 & 1.59576 & 1.59576 \\ 0 & 0 & 0 & -2.12769 & 0 & 2.12769 \\ 0 & 0 & 0 & 1.24115 & -1.59576 & 1.24115 \end{bmatrix}.$$

There is a relationship between the second Chebyshev wavelet and the block pulse function:

$$\Psi(t) = \Phi B_{m'}(t). \tag{4}$$

If I^α is the fractional-order integration operator of the second Chebyshev wavelets, one can achieve [17,24]:

$$I^\alpha \Psi(t) \approx P^\alpha \Psi(t), \quad \text{with} \quad P^\alpha = \Phi F^\alpha \Phi^{-1}, \tag{5}$$

where P^α is named as the operational matrix of fractional-order integration of the second Chebyshev wavelet. For example,

$$P^{0.5} = \begin{bmatrix} 0.5365 & 0.1575 & -0.0249 & 0.4366 & -0.0754 & 0.0214 \\ 0.0191 & -0.0449 & 0.0858 & 0.1287 & 0.2242 & -0.2105 \\ 0.0512 & -0.0470 & 0.1604 & 0.0948 & -0.0253 & 0.0100 \\ 0 & 0 & 0 & 0.5365 & 0.1575 & -0.0249 \\ 0 & 0 & 0 & -0.2105 & 0.2242 & 0.1287 \\ 0 & 0 & 0 & 0.0512 & -0.0470 & 0.1604 \end{bmatrix}.$$

3.2. Function Approximation

Using the second Chebyshev wavelet, each function in L^2 can be approximated using the following lemma.

Lemma 2. *Any function $f \in L^2([0,1])$ can be expanded into the second Chebyshev wavelet as [16,27,28]:*

$$f(t) = \sum_{n=1}^{\infty} \sum_{m \in \mathbb{Z}} c_{nm} \psi_{nm}(t), \tag{6}$$

where
$$c_{nm} = \langle f(t), \psi_{nm}(t) \rangle_{\omega_n} = \int_0^1 \psi_{nm}(t) f(t) \omega_n(t) dt.$$

If Equation (6) is truncated, then with Equation (3)
$$f(t) \approx \sum_{n=1}^{2^{k-1}} \sum_{m=0}^{M-1} c_{nm} \psi_{nm}(t) = C^T \Psi(t).$$

Remark 1. *Let* $X \in L^2([0,1])$, *so*
$$I^\alpha X(t) \approx I^\alpha C^T \Psi(t) = C^T I^\alpha \Psi(t) = C^T P^\alpha \Psi(t). \tag{7}$$

4. Method Analysis

For solving the system (1), without reducing the generality of the equations under consideration, we assume that the initial conditions are zero, and we approximate $D^\alpha y_i(t)$, $f_i(t)$, and $k_i(t,s)$ for $i = 1, 2$ in terms of the second Chebyshev wavelet as follows:
$$D^{\alpha_i} y_i(t) \simeq C_i^T \Psi(t), \qquad f_i(t) \simeq F_i^T \Psi(t), \qquad k_i(t,s) \simeq \Psi^T(t) K_i \Psi(s). \tag{8}$$

From Equations (2), (7) and (8), we obtain
$$y_i(t) = I^{\alpha_i} D^{\alpha_i} y_i(t) \simeq C_i^T P^{\alpha_i} \Psi(t). \tag{9}$$

Thus,
$$\begin{aligned}
\int_0^t \frac{y_i(s)}{(t-s)^{\beta_i}} ds &= C_i^T P^{\alpha_i} \int_0^t \frac{\Psi(s)}{(t-s)^{\beta_i}} ds \\
&= C_i^T P^{\alpha_i} \Gamma(1-\beta_i) I^{1-\beta_i} \Psi(t) \\
&= \Gamma(1-\beta_i) C_i^T P^{\alpha_i} P^{1-\beta_i} \Psi(t),
\end{aligned} \tag{10}$$

and from Equation (8) and $\int_0^1 \Psi(s) \Psi(s)^T ds = D$, we have
$$\begin{aligned}
\int_0^1 k_i(t,s) y_i(s) ds &= \int_0^1 \Psi^T(t) K_i \Psi(s) \Psi(s)^T P^{\alpha_i T} C_i ds \\
&= \Psi^T(t) K_i \int_0^1 \Psi(s) \Psi(s)^T ds P^{\alpha_i T} C_i \\
&= \Psi^T(t) K_i D P^{\alpha_i T} C_i = C_i^T P^{\alpha_i} D^T K_i^T \Psi(t).
\end{aligned} \tag{11}$$

By substituting Equations (8)–(11) into (1), we get
$$\begin{cases} C_1^T \Psi(t) = \lambda_1 \Gamma(1-\beta_1) C_1^T P^{\alpha_1} P^{1-\beta_1} \Psi(t) + \lambda_2 C_2^T P^{\alpha_2} D^T K_1^T \Psi(t) + F_1^T \Psi(t), \\ C_2^T \Psi(t) = \lambda_3 \Gamma(1-\beta_2) C_1^T P^{\alpha_1} P^{1-\beta_2} \Psi(t) + \lambda_4 C_2^T P^{\alpha_2} D^T K_2^T \Psi(t) + F_2^T \Psi(t), \end{cases} \tag{12}$$

and we obtain
$$\begin{cases} C_1^T = \lambda_1 \Gamma(1-\beta_1) C_1^T P^{\alpha_1} P^{1-\beta_1} + \lambda_2 C_2^T P^{\alpha_2} D^T K_1^T + F_1^T, \\ C_2^T = \lambda_3 \Gamma(1-\beta_2) C_1^T P^{\alpha_1} P^{1-\beta_2} + \lambda_4 C_2^T P^{\alpha_2} D^T K_2^T + F_2^T. \end{cases} \tag{13}$$

By solving system (13), one can get C_1 and C_2. Then substituting them into (9), the unknown solutions can be obtained.

5. Error Analysis

In this section, we present an error estimation for the system of Equation (1).

Theorem 1. Let $\hat{y}_1(t)$ and $\hat{y}_2(t)$ be the approximations of $y_1(t)$ and $y_2(t)$, obtained by the second Chebyshev wavelets basis (6), as the solutions of system (1) with $0 < \beta_1, \beta_2 < 1/2$. Assume also that there is a pair of constants k_1 and k_2 such that

$$\|k_i(t,s)\|_2 = <k_i(t,s), k_i(t,s)>^{\frac{1}{2}} = \left(\int_0^1 \int_0^1 |k_i(t,s)|^2 dt ds\right)^{\frac{1}{2}} \le k_i, \ i = 1,2.$$

If
$$|\lambda_1| < \Gamma(2+\alpha_1)\sqrt{1-2\beta_1},$$
$$k_2|\lambda_4| < \Gamma(1+\alpha_2),$$
$$\frac{|\lambda_1|}{(1+\alpha_1)\sqrt{1-2\beta_1}} + \frac{k_1|\lambda_2\lambda_3|}{(1+\alpha_2)\sqrt{1-2\beta_2}(\Gamma(1+\alpha_2) - k_2|\lambda_4|)} < \Gamma(1+\alpha_1),$$

and
$$\frac{k_1|\lambda_2\lambda_3|(1+\alpha_1)\sqrt{1-2\beta_1}}{(1+\alpha_2)\sqrt{1-2\beta_2}(\Gamma(2+\alpha_1)\sqrt{1-2\beta_1} - |\lambda_1|)} + k_2|\lambda_4| < \Gamma(1+\alpha_2),$$

then the approximate solutions of system (1) converge to the exact solutions with respect to L_2 norm.

Proof. Compute, by the assumptions,

$$\|y_1 - \hat{y}_1\|_2 = \left\| I^{\alpha_1}\left[\lambda_1 \int_0^t \frac{y_1(s) - \hat{y}_1(s)}{(t-s)^{\beta_1}} ds + \lambda_2 \int_0^1 k_1(t,s)(y_2(s) - \hat{y}_2(s)) ds\right]\right\|_2,$$

and

$$\|y_2 - \hat{y}_2\|_2 = \left\| I^{\alpha_2}\left[\lambda_3 \int_0^t \frac{y_1(s) - \hat{y}_1(s)}{(t-s)^{\beta_2}} ds + \lambda_4 \int_0^1 k_2(t,s)(y_2(s) - \hat{y}_2(s)) ds\right]\right\|_2.$$

Using the triangular inequality, we get

$$\|y_1 - \hat{y}_1\|_2 \le I^{\alpha_1}\left[|\lambda_1| \int_0^t \|(t-s)^{-\beta_1}\|_2 \|y_1 - \hat{y}_1\|_2 ds + |\lambda_2| \int_0^1 \|k_1(t,s)\|_2 \|y_2 - \hat{y}_2\|_2 ds\right],$$

and

$$\|y_2 - \hat{y}_2\|_2 \le I^{\alpha_2}\left[|\lambda_3| \int_0^t \|(t-s)^{-\beta_2}\|_2 \|y_1 - \hat{y}_1\|_2 ds + |\lambda_4| \int_0^1 \|k_2(t,s)\|_2 \|y_2 - \hat{y}_2\|_2 ds\right].$$

Furthermore, for $0 < \beta_1, \beta_2 < 1/2$ and $0 \le t \le 1$, we have

$$\|(t-s)^{-\beta_1}\|_2 = \left(\int_0^1 \frac{1}{(t-s)^{2\beta_1}} ds\right)^{\frac{1}{2}} \le \frac{1}{\sqrt{1-2\beta_1}},$$

$$\|(t-s)^{-\beta_2}\|_2 = \left(\int_0^1 \frac{1}{(t-s)^{2\beta_2}} ds\right)^{\frac{1}{2}} \le \frac{1}{\sqrt{1-2\beta_2}},$$

and as a result, we obtain

$$\|y_1 - \hat{y}_1\|_2 \le \frac{|\lambda_1|}{\Gamma(2+\alpha_1)\sqrt{1-2\beta_1}} \|y_1 - \hat{y}_1\|_2 + \frac{k_1|\lambda_2|}{\Gamma(1+\alpha_1)} \|y_2 - \hat{y}_2\|_2,$$

and

$$\|y_2 - \hat{y}_2\|_2 \leq \frac{|\lambda_3|}{\Gamma(2+\alpha_2)\sqrt{1-2\beta_2}}\|y_1 - \hat{y}_1\|_2 + \frac{k_2|\lambda_4|}{\Gamma(1+\alpha_2)}\|y_2 - \hat{y}_2\|_2.$$

Now, we denote

$$e_i(t) = \|y_i(t) - \hat{y}_i(t)\|_2, \quad i = 1, 2,$$

so,

$$e_1(t) \leq \frac{|\lambda_1|}{\Gamma(2+\alpha_1)\sqrt{1-2\beta_1}}e_1(t) + \frac{k_1|\lambda_2|}{\Gamma(1+\alpha_1)}e_2(t), \tag{14}$$

and consequently by the assumption $|\lambda_1| < \Gamma(2+\alpha_1)\sqrt{1-2\beta_1}$, one can conclude

$$e_1(t) \leq \frac{\frac{k_1|\lambda_2|}{\Gamma(1+\alpha_1)}}{1 - \frac{|\lambda_1|}{\Gamma(2+\alpha_1)\sqrt{1-2\beta_1}}} e_2(t). \tag{15}$$

We have similar relations for the second approximation, i.e.,

$$e_2(t) \leq \frac{|\lambda_3|}{\Gamma(2+\alpha_2)\sqrt{1-2\beta_2}}e_1(t) + \frac{k_2|\lambda_4|}{\Gamma(1+\alpha_2)}e_2(t), \tag{16}$$

and attention to assumption $k_2|\lambda_4| < \Gamma(1+\alpha_2)$ leads to

$$e_2(t) \leq \frac{\frac{|\lambda_3|}{\Gamma(2+\alpha_2)\sqrt{1-2\beta_2}}}{1 - \frac{k_2|\lambda_4|}{\Gamma(1+\alpha_2)}} e_1(t). \tag{17}$$

Substituting (17) into (14), and (15) into (16), we have

$$(1-\varepsilon_1)e_1(t) \leq 0, \quad (1-\varepsilon_2)e_2(t) \leq 0, \tag{18}$$

where

$$\varepsilon_1 = \frac{|\lambda_1|}{\Gamma(2+\alpha_1)\sqrt{1-2\beta_1}} + \frac{k_1|\lambda_2|}{\Gamma(1+\alpha_1)}\left(\frac{\frac{|\lambda_3|}{\Gamma(2+\alpha_2)\sqrt{1-2\beta_2}}}{1 - \frac{k_2|\lambda_4|}{\Gamma(1+\alpha_2)}}\right),$$

and

$$\varepsilon_2 = \frac{|\lambda_3|}{\Gamma(2+\alpha_2)\sqrt{1-2\beta_2}}\left[\frac{\frac{k_1|\lambda_2|}{\Gamma(1+\alpha_1)}}{(1 - \frac{|\lambda_1|}{\Gamma(2+\alpha_1)\sqrt{1-2\beta_1}})}\right] + \frac{k_2|\lambda_4|}{\Gamma(1+\alpha_2)}.$$

On the other hand, according to the assumptions of this theorem, we get $1 - \varepsilon_1 > 0$ and $1 - \varepsilon_2 > 0$; also, we know that $e_1(t) \geq 0$ and $e_2(t) \geq 0$. Therefore, the relations in (18) are satisfied only for $e_1(t), e_2(t) = 0$. The conditions of this theorem are sufficient to find the appropriate approximate solution of system (1), but they can be met in certain cases rarely. Furthermore, according to the relation (6) and the expressed discretization, the approximation of the solution is done in some node points on $[0, 1]$. In other words, according to Lemma 2, it is possible to get a suitable approximation so that $e_1(t) \to 0$ and $e_2(t) \to 0$ by the constant consideration of m and when $k \to \infty$, even if the conditions of Theorem 1 do not hold. □

6. Numerical Example

To demonstrate the efficiency of our proposed method, we consider the following example.

Example 1. Consider the system of fractional-order Volterra–Fredholm integro-differential equations with a weakly singular kernel:

$$\begin{cases} D^{0.4}y_1(t) = \int_0^t \frac{y_1(s)}{(t-s)^{0.5}} ds + 2\int_0^1 sty_2(s)ds + f_1(t), \\ D^{0.5}y_2(t) = -0.5\int_0^t \frac{y_1(s)}{(t-s)^{0.4}} ds + \int_0^1 (t+s)y_2(s)ds + f_2(t), \end{cases} \quad y_1(0) = 0, y_2(0) = 0,$$

where

$$f_1(t) = \frac{1}{\Gamma(1.6)} t^{\frac{6}{10}} - \frac{2}{3} t - \frac{4}{3} t^{\frac{3}{4}}, \quad f_2(t) = -\frac{1}{\Gamma(1.5)} t^{\frac{1}{2}} + \frac{1}{2} t + \frac{25}{48} t^{\frac{8}{5}} + \frac{1}{3}.$$

The exact solutions of this system are $y_1(t) = t$ and $y_2(t) = -t$. The absolute errors of $y_1(t)$ and $y_2(t)$ for different values of t are listed in Tables 1 and 2.

Figures 2–5 display the results of comparing the errors of different approximations by the second Chebyshev wavelets method for various values of k and M. Furthermore, Table 3 shows the execution time.

Table 1. Absolute error of $y_1(t)$ for $M = 3$ and $k = 4, 5, 6$ (Example 1).

t	$M = 3, k = 4$	$M = 3, k = 5$	$M = 3, k = 6$
0.1	1.5733×10^{-3}	5.3788×10^{-4}	1.7841×10^{-4}
0.2	8.1154×10^{-4}	2.6760×10^{-4}	8.9690×10^{-5}
0.3	1.7912×10^{-4}	5.1094×10^{-5}	1.4128×10^{-5}
0.4	1.4759×10^{-3}	4.6674×10^{-4}	1.4909×10^{-4}
0.5	3.1524×10^{-3}	1.0029×10^{-3}	3.2290×10^{-4}
0.6	5.2736×10^{-3}	1.6820×10^{-3}	5.4294×10^{-4}
0.7	7.9262×10^{-3}	2.5301×10^{-3}	8.1767×10^{-4}
0.8	1.1204×10^{-2}	3.5784×10^{-3}	1.1571×10^{-3}
0.9	1.5225×10^{-2}	4.8636×10^{-3}	1.5733×10^{-3}

Table 2. Absolute error of $y_2(t)$ for $M = 3$ and $k = 4, 5, 6$ (Example 1).

t	$M = 3, k = 4$	$M = 3, k = 5$	$M = 3, k = 6$
0.1	9.1898×10^{-3}	2.9619×10^{-3}	9.6013×10^{-4}
0.2	1.2525×10^{-2}	4.0140×10^{-3}	1.3022×10^{-3}
0.3	1.5780×10^{-2}	5.0567×10^{-3}	1.6400×10^{-3}
0.4	1.9240×10^{-2}	6.1633×10^{-3}	1.9983×10^{-3}
0.5	2.3036×10^{-2}	7.3765×10^{-3}	2.3910×10^{-3}
0.6	2.7270×10^{-2}	8.7311×10^{-3}	2.8296×10^{-3}
0.7	3.2061×10^{-2}	1.0263×10^{-2}	3.3254×10^{-3}
0.8	3.7524×10^{-2}	1.2009×10^{-2}	3.8909×10^{-3}
0.9	4.3800×10^{-2}	1.4015×10^{-2}	4.5404×10^{-3}

Table 3. Run times of Example 1.

Values of M, k	Run Time (s)
$M = 3, k = 4$	0.78
$M = 3, k = 5$	1.70
$M = 3, k = 6$	6.55

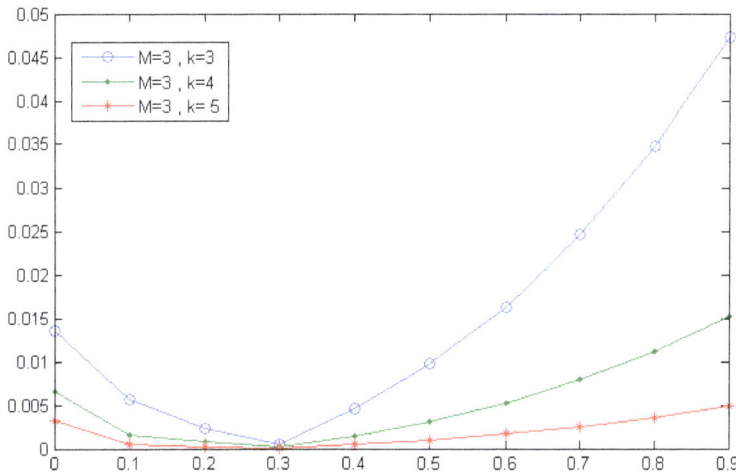

Figure 2. Absolute error of $y_1(t)$ for $M = 3$ and $k = 3, 4, 5$ (Example 1).

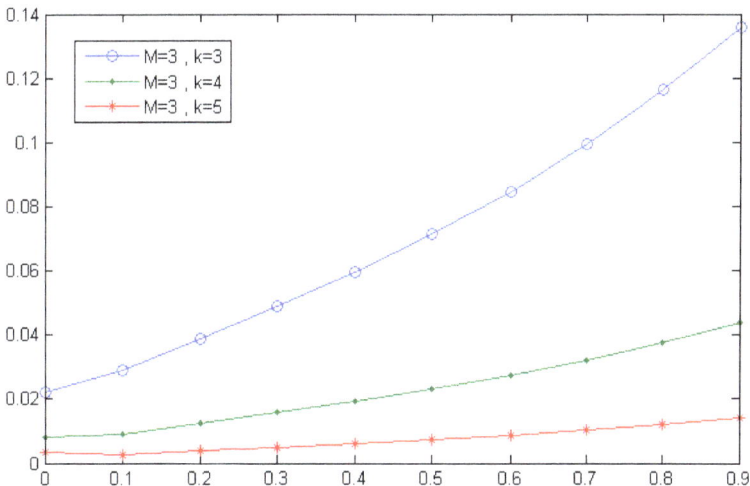

Figure 3. Absolute error of $y_2(t)$ for $M = 3$ and $k = 3, 4, 5$ (Example 1).

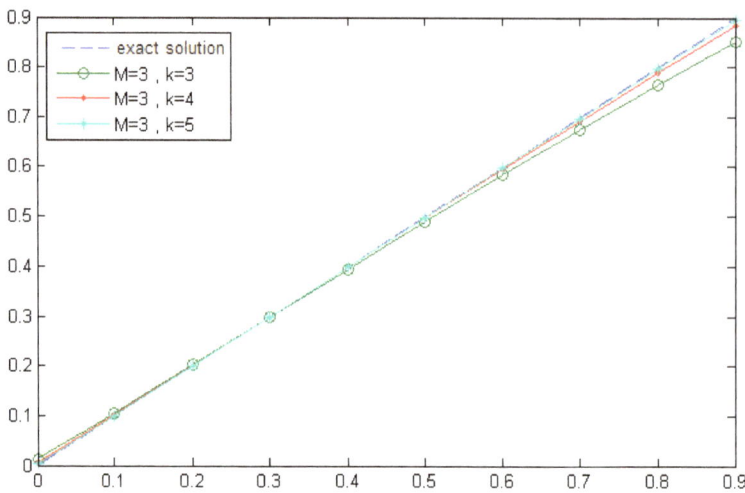

Figure 4. Exact and approximate solutions of $y_1(t)$ for $M = 3$ and $k = 3, 4, 5$ (Example 1).

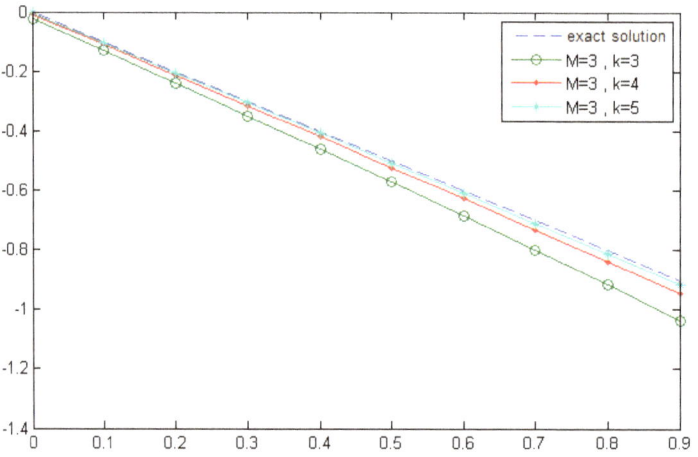

Figure 5. Comparison between the exact and numerical solutions of $y_2(t)$ for $M = 3$ and $k = 3, 4, 5$ (Example 1).

7. Conclusions

In this paper, a numerical algorithm using the second Chebyshev wavelet was proposed for a system of fractional-order Volterra–Fredholm integro-differential equations with weakly singular kernels in the Volterra part. Applying the properties of the second Chebyshev wavelet, we transformed the main system into an algebraic system of equations with a sparse coefficient matrix. By solving this system, an approximate solution was obtained for the system of fractional integral equations. Furthermore, the error analysis of the proposed approach was presented.

Author Contributions: Conceptualization, E.B., L.T. and K.N.; methodology, E.B., L.T. and A.J.; validation, K.N. and A.J.; formal analysis, E.B., L.T. and K.N.; investigation, L.T., K.N. and A.J.; writing—original draft preparation, E.B. and K.N.; writing—review and editing, L.T. and A.J. All authors have read and agreed to the published version of the manuscript.

Funding: This research received no external funding.

Institutional Review Board Statement: Not applicable.

Informed Consent Statement: Not applicable.

Data Availability Statement: All the data is present within the manuscript.

Conflicts of Interest: The authors declare no conflict of interest.

References

1. Baleanu, D.; Sajjadi, S.S.; Jajarmi, A.; Defterli, O. On a nonlinear dynamical system with both chaotic and non-chaotic behaviours: A new fractional analysis and control. *Adv. Differ. Equ.* **2021**, *2021*, 234. [CrossRef]
2. Jajarmi, A.; Baleanu, D. On the fractional optimal control problems with a general derivative operator. *Asian J. Control* **2021**, *23*, 1062–1071. [CrossRef]
3. Baleanu, D.; Sajjadi, S.S.; Asad, J.H.; Jajarmi, A.; Estiri, E. Hyperchaotic behaviours, optimal control, and synchronization of a nonautonomous cardiac conduction system. *Adv. Differ. Equ.* **2021**, *2021*, 157. [CrossRef]
4. Yi, M.; Huang, J. CAS wavelet method for solving the fractional integro-differential equation with a weakly singular kernel. *Int. J. Pure Appl. Math.* **2015**, *92*, 1715–1728. [CrossRef]
5. Wang, Y.; Zhu, L. SCW method for solving the fractional integro-differential equations with a weakly singular kernel. *Appl. Math. Comput.* **2016**, *275*, 72–80. [CrossRef]
6. Sahu, P.K.; Saha Ray, S. Legendre wavelets operational method for the numerical solutions of nonlinear Volterra integro-differential equations system. *Appl. Math. Comput.* **2015**, *256*, 715–723. [CrossRef]
7. Baleanu, D.; Jajarmi, A.; Mohammadi, H.; Rezapour, S. A new study on the mathematical modelling of human liver with Caputo-Fabrizio fractional derivative. *Chaos Soliton. Fract.* **2020**, *134*, 109705. [CrossRef]
8. Baleanu, D.; Jajarmi, A.; Asad, J.H.; Blaszczyk, T. The motion of a bead sliding on a wire in fractional sense. *Acta Phys. Pol. A* **2017**, *131*, 1561–1564. [CrossRef]
9. Baleanu, D.; Sajjadi, S.S.; Jajarmi, A.; Defterli, O.; Asad, J.H. The fractional dynamics of a linear triatomic molecule. *Rom. Rep. Phys.* **2021**, *73*, 105.
10. Baleanu, D.; Ghanbari, B.; Asad, J.H.; Jajarmi, A.; Mohammadi Pirouz, H. Planar system-masses in an equilateral triangle: Numerical study within fractional calculus. *CMES-Comput. Model. Eng. Sci.* **2020**, *124*, 953–968. [CrossRef]
11. Sahu, P.K.; Saha Ray, S. Hybrid Legendre Block-Pulse functions for the numerical solutions of system of nonlinear Fredholm–Hammerstein integral equations. *Appl. Math. Comput.* **2015**, *270*, 871–878. [CrossRef]
12. Yüzbaşı, S. Numerical solutions of system of linear Fredholm-Volterra integro-differential equations by the Bessel collocation method and error estimation. *Appl. Math. Comput.* **2015**, *250*, 320–338. [CrossRef]
13. Deif, A.S.; Grace, R.S. Iterative refinement for a system of linear integro-differential equations of fractional type. *J. Comput. Appl. Math.* **2016**, *294*, 138–150. [CrossRef]
14. Xie, J.; Yi, M. Numerical research of nonlinear system of fractional Volterra-Fredholm integral-differential equations via Block-Pulse functions and error analysis. *J. Comput. Appl. Math.* **2019**, *345*, 159–167. [CrossRef]
15. Saemi, F.; Ebrahimi, H.; Shafiee, M. An effective scheme for solving system of fractional Volterra–Fredholm integro-differential equations based on the Müntz–Legendre wavelets. *J. Comput. Appl. Math.* **2020**, *374*, 112–773. [CrossRef]
16. Lal, S.; Sharma, R.P. Approximation of function belonging to generalized Hölder's class by first and second kind Chebyshev wavelets and their applications in the solutions of Abel's integral equations. *Arab. J. Math.* **2021**, *10*, 157–174. [CrossRef]
17. Zhu, L.; Wang, Y. Numerical solutions of Volterra integral equation with weakly singular kernel using SCW method. *Appl. Math. Comput.* **2015**, *260*, 63–70. [CrossRef]
18. Zhang, Z. Legendre wavelets method for the numerical solution of fractional integro-differential equations with weakly singular kernel. *Appl. Math. Model.* **2016**, *40*, 3422–3437.
19. Maleknejad, K.; Nouri, K.; Torkzadeh, L. Operational matrix of fractional integration based on the shifted second kind Chebyshev Polynomials for solving fractional differential equationss. *Mediterr. J. Math.* **2016**, *13*, 1377–1390. [CrossRef]
20. Zhu, L.; Fan, Q. Solving fractional nonlinear Fredholm integro-differential equations by the second kind Chebyshev wavelet. *Commun. Nonlinear Sci. Numer. Simul.* **2012**, *17*, 2333–2341. [CrossRef]
21. Zhou, F.; Xu, X. Numerical solution of the convection diffusion equations by the second kind Chebyshev wavelets. *Appl. Math. Comput.* **2014**, *247*, 353–367. [CrossRef]
22. Wang, Y.; Fan, Q. The second kind Chebyshev wavelet method for solving fractional differential equations. *Appl. Math. Comput.* **2012**, *218*, 8592–8601. [CrossRef]
23. Manchanda, P.; Rani, M. second kind Chebyshev wavelet method for solving system of linear differential equations. *Int. J. Pure Appl. Math.* **2017**, *114*, 91–104. [CrossRef]
24. Zhu, L.; Wang, Y. Solving fractional partial differential equations by using the second Chebyshev wavelet operational matrix method. *Nonlinear. Dyn.* **2017**, *89*, 1915–1925. [CrossRef]
25. Tavassoli Kajani, M.; Hadi Vencheh, A.; Ghasemi, M. The Chebyshev wavelets operational matrix of integration and product operation matrix. *Int. J. Comput. Math.* **2009**, *86*, 1118–1125. [CrossRef]

26. Zhou, F.; Xu, X.; Zhang, X. Numerical integration method for triple integrals using the second kind Chebyshev wavelets and Gauss–Legendre quadrature. *Comp. Appl. Math.* **2018**, *37*, 3027–3052. [CrossRef]
27. Yi, M.; Ma, L.; Wang, L. An efficient method based on the second kind Chebyshev wavelets for solving variable-order fractional convection diffusion equations. *Int. J. Comput. Math.* **2018**, *95*, 1973–1991. [CrossRef]
28. Negarchi, N.; Nouri, K. Numerical solution of Volterra—Fredholm integral equations using the collocation method based on a special form of the Müntz—Legendre polynomials. *J. Comput. Appl. Math.* **2018**, *344*, 15–24. [CrossRef]

MDPI
St. Alban-Anlage 66
4052 Basel
Switzerland
Tel. +41 61 683 77 34
Fax +41 61 302 89 18
www.mdpi.com

Fractal and Fractional Editorial Office
E-mail: fractalfract@mdpi.com
www.mdpi.com/journal/fractalfract

www.ingramcontent.com/pod-product-compliance
Lightning Source LLC
LaVergne TN
LVHW070626100526
838202LV00012B/734